November 1989

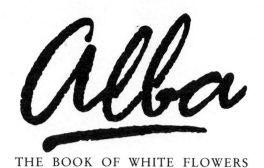

THE BOOK OF WHITE FLOWERS

To Mum and Dad

Alba

THE BOOK OF WHITE FLOWERS

Deni Bown

Timber Press
Portland, Oregon

First published in Great Britain by Unwin Hyman,
an imprint of Unwin Hyman Limited, 1989

UNWIN HYMAN LIMITED
15–17 Broadwick Street
London W1V 1FP

Allen & Unwin Australia Pty Ltd
8 Napier Street
North Sydney, NSW 2060
Australia

Allen & Unwin with the Port Nicholson Press
Compusales Building
75 Ghunzee Street
Wellington
New Zealand

British Library Cataloguing in Publication Data

Bown, Deni
    Alba: the book of white flowers.
1. Flowering plants. Flowers
I. Title
582′.0463

ISBN 0–04–440375–5

Designed by Julian Holland
Typeset by Nene Phototypesetters Ltd, Northampton
Printed and bound in Great Britain by Butler and Tanner Ltd, Frome

First published in North America in 1989 by
Timber Press, Inc.
9999 S.W. Wilshire
Portland, Oregon 97225

ISBN 0–88192–157–2

Acknowledgements

The author and publishers would like to thank the following for their
kind permission to use their photographs and illustrations on the pages
listed. Slough Corporation for their assistance and permission to
reproduce the Slough Coat of Arms, 22; The Royal Library, Windsor
Castle for *Lilium Candidum* by Leonardo da Vinci (© 1989 Her Majesty
the Queen), 31; the Tate Gallery, London for *Ecce Ancilla Domine* by
Dante Gabriel Rossetti, 32, and for *The Resurrection* by Sir Stanley
Spencer, 37; Academy Editions, London for the illustration of the
nativity by Grasset, reproduced from *Berthon and Grasset* by Victor
Arwas, 35; Emilia Guzzy de Galvez for *Nude with Calla Lilies* by Diego
Rivera, 36; The Royal Photographic Society, Bath for *Bringing home the
May* by Henry Peach Robinson, 43; and Andy Weber and Tharpa
Publications for the drawing of Buddha, 47.

Contents

Sunlight made visible
the whole length of the sky,
movement of wind,
leaf, flower, all six colours
on tree, bush and creeper:
                    all this
is the day's worship.

The light of moon, star and fire,
lightnings and all things
that go by the name of light
are the night's worship.

                    Night and day
                    in your worship
                    I forget myself

O lord white as jasmine.

A lyrical hymn to Shiva by
Mahadeviyakka, twelfth-century
mystic and poetess (translated from the
medieval *Kannada* by A. K. Ramanujan).

Introduction

Flowers are the inspiration and aspiration of our lives. As Iris Murdoch wrote, 'people from another planet without flowers would think we must be mad with joy the whole time to have such things about us' – and indeed, some of us are! Yet on the whole, their influence is more pervasive than dramatic, for in spite of their beauty and the rarity of individual species, they are also commonplace. Everyday countless millions unfold in rhythmic response to the earth's fertility, and their images have shaped art, design and expression through the centuries. Even those who live far from the countryside or have no garden are never very far from flowers, as floral themes and motifs are part of daily life, depicted on a diversity of household items such as fabrics, toiletries and greetings cards. They are figuratively present in our language too. We talk of 'budding talent', of someone 'blooming' or 'blossoming', of 'the flower of youth' and 'the flowering of a civilisation'. For the plant, they are simply the means to sexual fulfilment but we are moved by their beauty to a far wider expression of hope, whether in our saddest or most joyful moments.

Everyone has a favourite flower or one which holds a particular association, but of them all, none can compare with white flowers for the degree with which they are loved, hated or feared – which is odd since white, of all flower colours, would appear to be the most neutral. In theory it should verge on the insipid, but in practice it provides contrast and is an extreme. No doubt if there were such things as black flowers, they would receive their share of attention too, but the fact is that there are remarkably few flowers which even remotely approach black in their coloration, whereas many are virtually pure white.

Whatever its definition according to the laws of physics, white is psychologically more, not less than a colour, and in most cultures has particular significance. On the one hand, its luminosity suggests something more spirit than flesh, an unsullied freshness and purity which signifies innocence, chastity and other-worldliness. Conversely, its pallor is reminiscent of fear and death, of the hygienic and the sick, the shocked and the ghostly.

We not only enjoy the beauty of flowers and exploit their decorative potential but also use them for communication. The floral tribute – saying it with flowers – is an age-old way of commemorating a wide range of occasions, from birth and birthdays, death, illness and tragedies to love, marriage, grati-tude, apologies and respect. Consequently, they are closely associated with some of our most moving and emotional experiences. The choice may vary according to culture and climate, and fashions change, but white flowers are enduringly popular the world over for their symbolism of higher values. Ironically, they are also among the most unpopular of flowers, for many people find them inseparable from emotionally charged events which are painful to recall.

Highly scented white flowers bear this burden of ambiguity more than any others. At weddings they symbolise virginity and fidelity, and above all bring delight, but their role at funerals is to represent death and the afterlife, as well as originally serving the practical purpose of disguising the odour of decay. Our memories of happy events are often fleeting, whereas those involving suffering tend to surface time and time again throughout our lives as a result of our attempts to repress them. Not surprisingly then, those who have been in mourning tend to associate such flowers with grief from then on, which could explain why so many scented white flowers have passed into folklore as blooms of ill-omen.

The memory of an intense experience or certain period of one's life is often bound up with a scent which was present at the time, and years later may be triggered by it. To this day I am repelled and nauseated by the smell of flowering privet as I lived near an overgrown hedge of it when suffering from morning sickness in two consecutive pregnancies. It is certain-ly my *bête noire* of scented white flowers (though the term 'black beast' seems rather bizarre in the context). On the other hand, I am besotted by lilies, which always remind me of my grandfather who used to grow them in a greenhouse for the cut-flower trade. As a child, I was given the job of removing the anthers (whose brown pollen would ruin the blooms in transit) and as I sat intent on my task in the silence of white trumpets, I was aware of breathing in the humid scented air that the lilies seemed to be breathing out – an experience which is still one of my greatest pleasures to recall.

If flowers which are normally white can evoke strong feelings, then those with rare white forms can be expected to cause even more passionate adoration or revulsion. White-flowered forms are given recognition by the Latin word *alba* (or *albus* or *album*, according to the grammatical gender of the preceding name), meaning white, which is added on to the species' name: as in *Primula denticulata* 'Alba'. Some gar-deners cannot abide white versions of flowers, finding them

Primula denticulata (drumstick primula) – the usual purple species is here growing alongside its white or 'alba' form.

anaemic by comparison with the normal colour and somehow not quite right. Others find them irresistible, as they do all white flowers (and it is for these perceptive souls that this book is written), while most would acknowledge that the colour variation gives scope for a wider range of uses.

'Alba' forms are not, however, as innocent as their whiteness would suggest. The whole subject of albinism is charged with subconscious hopes and fears. Scientists may be able to find reasons for such phenomena, but albinos – whether plant, animal or human – still fascinate as they did long before the workings of pigments and genes were understood. When unexpectedly encountered, they stop you in your tracks and inspire awe – that strange mixture of alarm and admiration which gives rise to stories of legendary dimensions.

In many countries, albinos of our own species are feared and ostracised. The albino singer Salif Keita was born into the Mali royal family but became an expatriate wandering minstrel,

rejected by relatives and countrymen alike. Albino animals, however, are often revered. White elephants are regarded as sacred in parts of southern Asia (though they have also presumably given westerners the euphemism for something rare and valuable which, as far as material considerations go, is also completely useless or impossibly expensive to keep). It seems we are compulsively drawn to creatures such as Moby Dick, the white whale, the hunt for whom became a personal odyssey. Even creatures which are normally white become imbued with this mystique of power. It is no coincidence that the dreaded killer of the film *Jaws* is a great white shark, and in Peter Matthiessen's book *The Snow Leopard*, that palest and most beautiful of spotted cats takes on a similar aura of myth – so much so that one author (David Helton) wrote that 'anybody

Lilium longiflorum (Easter lily) – a classic (and classy) white lily: ancient symbol of purity and modern designer image of style.

who was ever killed by one would certainly go straight to Paradise'. White birds take on symbolic meanings too – storks are associated with birth, cranes with happiness, and white doves are an almost universal emblem of peace. In Samuel Taylor Coleridge's poem *The Ancient Mariner*, an albatross inspires a powerful tale of good and evil. The seafarer shoots the great white bird and his shipmates hang the corpse around his neck as a constant reminder that guilt and despair befall those who are so insensitive to life as to destroy their very source of hope and inspiration.

One of the strangest examples of albinism is in creatures which live in the perpetual darkness of caves. For years I kept a blind cave fish in a tropical aquarium. It was that peculiar pinkish-white of albinos and had no eyes. Such fish are popular with aquarists and live happily with the other occupants of the tank. They are found all over the world, this particular one coming from a subterranean system of pools and streams in

Mexico. Many other kinds of animals have adapted to living in pitch blackness by becoming white and blind. They are also, on the whole, very long-lived because of their slow metabolic rate. No wonder Tolkien gripped the imagination with his tale of Gollum, the pale-eyed troglodyte of *The Hobbit* and *The Lord of the Rings*, who became a refugee from the world of light, 'wormed his way like a maggot into the heart of the hills' and lived by catching blind cave fish and anything else he could grasp and strangle with his long fingers. In this underworld, the dark is light enough.

White flowers, like white creatures – especially those which are not normally white – can have an air of unreality, a strangeness which jolts our expectations and gives cause for wonder. The plant hunters of the past, to whom we owe so many of our garden and greenhouse plants, were often expansive on the subject, their writings full of accounts of finds which held them spellbound: Reginald Farrer and the 'overpoweringly superb' crinkled goblets of *Paeonia moutan*; E. H. Wilson and his 'royal lady', *Lilium regale*; Benedict Roezl and the incredibly rare (if not unique) and beautiful *Cattleya skinneri* 'Alba', to mention but a few. Even now I am awestruck by the towering spires of the giant Himalayan lily (*Cardiocrinum giganteum*) and the handkerchief tree (*Davidia involucrata*) in full flower, so I can (just) imagine the ecstasy of their discoverers.

White flowers may on occasion be awe-inspiring – I think I would fall to my knees if I suddenly came across *Cardiocrinum giganteum* on a moonlit night – but our usual response is more casual. They are refreshing, relaxing, romantic – even contemplative. White gardens come into their own in the evening when we are most susceptible to tranquillity. The whiteness of the flowers intensifies as other colours become invisible, their alluring scents hang heavy in the still night air, and we are lulled into a meditative mood – Wordsworth's 'sense sublime of something far more deeply interfused'. And we call white flowers such names as 'Angel Wings', 'Summer Snow', 'L'Innocence', 'Flight of Cranes', 'Kingdom of the Moon', investing their beauty with our visions.

However fascinating the mystique of white flowers may be, there is a thoroughly practical side to them too. No serious gardener is unaware of their importance, even if white gardens as such are of no interest. White is the colour which displays others to their best advantage. It both brightens and relieves accompanying colours. Examples are legion. Yellow and lime green are rather too close for comfort as a colour scheme yet become a refreshing and sophisticated combination when white is added. Red brick walls are anathema to shades of pink, purple and crimson, and cry out for cascades of white or cream. Pastel shades can be a little limp on their own but crisp with a touch of white, and a splash of white makes bright colours even more dazzling. White flowers, especially alongside white-variegated foliage, lighten shady corners and give a feeling of coolness to sun-drenched conservatories and patios.

Gardeners are not alone in valuing white flowers. From the interior designer's point of view there is nothing to equal them for their unique ability to enhance and harmonise where other shades tend to clash or distract. And the importance of white flowers to florists goes without saying. They are still the mainstay for weddings and funerals, and, if anything, are increasing in popularity as flowers for romance, with white roses gaining fast on red as the floral offering for Valentine's Day – perhaps in subconscious recognition that 'faith, hope and chastity' are the new morality for the late eighties. The popularity of roses never seems to wane, but if I had to name the flower which more than any has captured the spirit of the present era, it would have to be *Lilium longiflorum*. Known unceremoniously as 'longi' in the trade, it constantly appears as the elegant *genius loci* in advertisements for fashionable decor, clothes and cosmetics. Equally effective, it makes a funereal appearance in videos and posters for the campaign against AIDS. What is it about the image of this flower that can sell anything from Wedgwood china to washing machines? It is pure white with a cool green tinge in its heart, has uncompromisingly clean lines, a waxy texture which is almost too perfect to be natural, and a sophisticated fragrance. The emphasis is definitely on style and superiority, but those who know it intimately will discover that beneath the flawless skin of this unreal beauty lies a vein of sensuousness and the eternal pulse of idealism.

DENI BOWN
Witney, January 1989

Botanic Alba – The Nature of White Flowers

White flowers are by no means unusual. They are found all over the world, though they are not evenly distributed – far northern regions have the most; deserts the fewest. The colour white is produced either by pigments or by the same optical illusion by which colourless ice crystals look white as snow. Flower colour is an important factor in attracting pollinators: bees see white flowers as blue-green; flies love clustered heads of tiny white flowers; moths and bats are drawn to pale nocturnal flowers which waft heavy scents into the night air. Some flowers, such as snowdrops, orange blossom and madonna lilies, are so invariably white that they are unthinkable in any other shade. Others may be white in the wild but have undergone so many changes in cultivation that the original colour is not necessarily the commonest: the peony (*Paeonia lactiflora*) being a good example. Then there are those whose normal colour is something other than white but which produce white-flowered offspring now and again. This is particularly common among species with flowers in the pink-purple-blue range – among bluebells and hardy geraniums, for instance. Lastly there are those for whom white flowers are one in a million (or greater). These are rarest in species with bright yellow flowers. Have you ever found a white dandelion?

Flower colour may be a matter of aesthetics for us but for the plant it has altogether more practical consequences. In some species, such as the madonna lily (*Lilium candidum*), the flowers are always white but many have another colour as the norm and white-flowered plants occur now and again. A good example is the bluebell (*Hyacinthoides non-scriptus*). Among the countless thousands which carpet the woodlands of northern Europe in the spring, there is almost always an odd white-flowered plant which stands out clearly in the haze of blue. White plants are also common in wild populations of foxgloves (*Digitalis purpurea*). Such deviations are referred to as mutations (or sports if only part of the plant – one branch of a tree, for example – is affected). The frequency of mutations is very variable according to the species: some regularly producing 'alba forms', as they are often called, and others seldom or rarely. White-flowered plants among wild primroses (*Primula vulgaris*) are, for instance, much less common than among bluebells.

As far as the plant is concerned, white flowers can have very different consequences, depending on whether it is the norm or a freak event. The usual pollinator may ignore a chance white-flowered plant, or another pollinator may be attracted and thus set the scene for changes which may, over many generations, result in the white-flowered plants differing sufficiently to be regarded as a new species. On the other hand, white-flowered forms are often weaker than their coloured counterparts and may not thrive or survive in the wild, though they do very well in cultivation where there is less competition and where the difficult task of reproduction is taken care of by the grower. Some even lack all pigments, including chlorophyll, and have white leaves too. Such complete albinos inevitably die at an early stage. Sometimes, though, a white-flowered form with variegated leaves may have just sufficient chlorophyll for photosynthesis to take place and, given the sheltered life of a garden plant, may succeed.

Human beings have always had a sharp eye for variation in the surrounding vegetation, nurturing superior forms and oddities which in many instances have proved of key importance in plant breeding. In some cases, white-flowered species have been introduced to cultivation, only to be eclipsed by colourful hybrids. We rarely see the wild single white *Paeonia lactiflora* but few gardens are without one of the double-forms which are now available in every shade of ruby, cerise and pastel pink. In contrast, white forms of species are much commoner in cultivation than in the wild, being treasured for their beauty, as well as for their possible role in breeding, and are consequently propagated with great care. It is well known among orchid breeders that the best miniature cymbidiums have *Cymbidium pumilum* 'Album' in their ancestry.

The colour white

Strictly speaking, white (and black) are not colours but the absence of colour. White sunlight is actually a mixture of red, orange, yellow, green, blue, indigo and violet, and it is split into these components when passed through a prism. (They are not really separate colours but a continuous change, as you can see in a rainbow.) The colours we see in the objects around us arise because they reflect certain parts of this

Digitalis purpurea – *white-flowered plants are quite common among wild foxgloves. White forms seldom come true from seed, so you must expect a proportion of coloured offspring from a white-flowered parent.*

The importance of light was grasped by human beings long before they had access to the explanations of physics. It became a symbol of the highest good, upon which life is dependent, and was manifest in a reverence for white-coloured things. Garments such as wedding dresses, christening robes, religious vestments, and shrouds are traditionally white, and many different white flowers and animals – lilies, roses and iris, doves, snow geese and storks, whales, elephants and the mythical unicorn are endowed with supernatural meaning, their whiteness being seen as the very incarnation of life-giving light.

White light is not, however, the same as white pigment – the substance used to create most white objects. If you mix the colours of the spectrum together, you get white light, but a mixture of pure pigments would produce black! The physical side to white is not always just a question of physics but may also involve chemistry and physiology. In practice, there is even more to white than this, for the perception of white as seen in flowers is also influenced by the plant's surroundings, and our own assumptions and psychology.

A white flower may be used as a symbol of simplicity, but in reality is very far from simple. White light in physics is an unchanging phenomenon but white in flowers is immensely variable. Colour charts for paint reveal a range of whites, from the dazzling to greyish and with every pastel tint, but the range is nevertheless limited compared with that of white flowers. Though many flowers are uniformly white, many are flushed or patterned with one or more other colours which change the tone of whiteness. In addition, a number of white flowers begin as differently coloured buds, changing colour as they age, either to a paler or darker shade or to a colour something other than white.

Another complication is that the whiteness of white flowers is greatly influenced by surrounding colours. When we look at white flowers, we instinctively make the necessary correction: a white rose in candle light is actually yellow, white bluebells under the green woodland canopy are more green than white and even the most pristine white lily blushes in the rays of the rising or setting sun, but nevertheless we perceive them as white. In a photograph, we make no such allowances and the actual colour takes precedence over the real colour. A glance at the photographs in this book will prove the point; indeed, one of the best ways of beginning to appreciate how variable is the colour (or absence of colour) we call white is to photograph white flowers and compare them. Our more objective perception reveals many different shades, both in the flowers themselves and in the influence that colour casts from their surroundings or lighting conditions has upon them.

spectrum: green leaves reflect green but absorb all the other colours. An object appears white because its surface reflects all or almost all the light reaching it, and conversely, something black absorbs the light. In biological terms, all life is dependent on light. Light transfers energy from one place to another and energises the surfaces which absorb it. It warms the earth and powers the reactions in the chlorophyll of green leaves, changing carbon dioxide and water into the food of living things. White objects reflect all or nearly all the light reaching them, passing it on to our eyes with little loss in intensity.

But colour casts apart, what causes the whiteness of white flowers? Colour in flowers is produced either by pigmentation or by physical effects (or both). The most common physical effect in white flowers is that of having colourless petals filled with air spaces which reflect the light. These effects are similar to the well-known phenomenon that water, itself colourless, looks white when frozen as snow because of the air spaces trapped by the crystals. The way light is reflected from white flowers depends on the texture of the petal. A gloxinia (*Sinningia speciosa*) has a velvety matt texture which is shown under the microscope to consist of extremely fine hair-like processes. A surface like this scatters light in all directions and the reflection is diffuse. In contrast, the outer cells of lesser celandine (*Ranunculus ficaria*) petals are filled with oil and backed by a layer of cells containing white starch. The result is a surface which is exceedingly smooth and mirror-like, causing a highly regular reflection giving a very shiny appearance. In the case of dazzling crystalline-white flowers, such as pure white African violets (*Saintpaulia ionantha*), the light is refracted, causing the flowers to look brighter than the surrounding light. What happens is that the cells are so arranged as to trap the light, bending and splitting the rays so that they bounce from cell to cell and cause internal reflection. If a flower is white for physical reasons, you can test this by squeezing the petal hard. Once the air is expelled, the limp petal you are left with will appear colourless.

The other cause of whiteness in flowers is pigmentation. There are two main categories of pigment: the ones in the cell sap being known as flavonoids; and the others contained in bodies called plastids which are situated within the cell. The latter include chlorophyll (the colouring matter present in

Saintpaulia hybrid – *whiteness can be caused by a number of different factors. The African violet's white petals are actually colourless and filled with air spaces which reflect light, as in snow.*

leaves, stalks and any greenish flower parts), which enables photosynthesis to take place, and carotenoids, which give various shades of red, orange, brown and yellow. More important for white flowers though are the flavonoids. Again there are two main kinds: anthocyanins (which cause red, blue and purple shades) and anthoxanthins (responsible for ivory, cream and yellows). (Many scarlets, oranges, browns and blackish shades are a combination of both.) The ones which are relevant to white flowers are the ivory anthoxanthins which produce ivory and cream shades of white. Apart from their importance in white flowers, they are a key component in plant breeding: though they have very little colour on their own, they act as co-pigments to increase blueness when they occur in conjunction with anthocyanins. Experimentally, the presence of anthoxanthins in white flowers can be demonstrated by the yellowing of petals exposed to ammonia vapours.

In some species and hybrids there are both ivory (creamy white) and pure white (anthoxanthin-free) cultivars – in antirrhinums, for example – which is a significant difference in varieties used for bedding schemes where colour contrasts and harmonies are so important. Pure white would be more effective than cream in combination with dark purple but a warmer white might look better with shades of blue or gold. In lupins, the flowers may be bi-coloured, with different pigments in the two halves of the flower. The cultivar 'Noble Maiden' has a white standard (upper petal) and cream keel (joined lower petals); in 'Joyce Henson' half of each flower is white, while the other half is blush-pink from the anthocyanins it contains.

Pigments are not always evenly distributed in the flower, as has just been mentioned in connection with lupins, and may undergo changes as the flower ages. The cell sap in a bud or newly opened flower tends to be acidic, producing a pink flush in white flowers which contain some anthocyanins. This fades as the flower ages and the sap becomes alkaline. Quite often the pigments occur in separate layers, making the front of a petal different in colour from the back, or giving patterned effects. A number of lilies are pure white inside the trumpet but are flushed maroon on the outside. Many white-flowered forms have anthocyanins present in coloured 'veins' (nectar guides) and/or in the anthers and stigmas, giving interesting and colourful detail when the flowers are looked at closely. The hardy geranium *G. clarkei* 'Kashmir White' has an exquisite feathering of mauve veins; from the centres of white fuchsias protrude delicate pink stamens; and the pure white zonal pelargonium 'Century White' has chalk-white single flowers with a scarlet dot of anthers in the centre. There are even different white forms of the same species: the cultivated white-flowered variant of *Geranium sylvaticum*, known as

G. sylvaticum 'Album', has pale foliage, pure white (anthocyanin-free) flowers, and comes true from seed; whereas wild albinos (termed *G. sylvaticum* forma *albiflorum*) have white petals but pink-coloured sepals, stamens and stigmas, and presumably do not come true from seed. These subtle variations repay close examination. In fact, white flowers are rarely 'dead' white!

The presence or absence and distribution of these pigments, their particular chemical structure and other factors (such as the degree of acidity or alkalinity) which affect the precise hue, are ultimately controlled genetically. Genetic changes take place in the course of cross-pollination but may have other causes, such as exposure to radiation, which may be difficult to determine. There are some rules of thumb about the inheritance of certain characteristics. For example, the presence of anthocyanins in, say, a purple-flowered species, is dependent on two different genes (A and B). The loss of either gene produces white-flowered offspring. A third gene (C) is responsible for colour modification (to pink, for example) but is only effective with A and B. The white-flowered form may carry gene A, B or C only, or a combination of AC or BC. It is obvious that if two white-flowered plants, one with A and the other with B, are crossed, their offspring will have anthocyanins and therefore purple flowers. To complicate matters, if white-flowered cultivars with the genetic combination of AC and BC are crossed, the third factor will also be present, giving a modification of the original purple colour (perhaps to pink) in the progeny. This explains why seed saved from some white-flowered forms does not necessarily produce another generation of white-flowered plants.

A book published in the nineteenth century (*The Colour of Flowers* by Grant Allen) proposed the theory that in the beginning all flowers were yellow. During the course of evolution, some progressed to white. A proportion of these then went on to become red or purple, and subsequently a few arrived at the most advanced colours: lilac, mauve and blue shades. In some cases, the advanced colour is apparent in the peripheral parts of the flower and its ancestral colour is retained toward the centre or base. According to this theory, if you examine a family of plants, you will find that its most primitive members have yellow flowers. However, reversion is common. Blue-flowered species commonly produce 'throwbacks' to pink or white and those with normally white flowers often revert to yellow (mainly to primrose or greenish-yellow, rather than to bright yellow). On the other hand, yellow-flowered species tend to be remarkably stable as they have no former colour. Whatever the truth of these observations, there is no doubt that *alba* forms are indeed quite common among most species with blue, red, purple and pink (i.e. anthocyanin-containing) flowers but are extremely rare in those whose flowers are bright yellow. The plantsman E. A. Bowles stated in *My Garden in Autumn and Winter* that 'every plant, except perhaps those with yellow flowers, of too primitive or conservative a plan to break out in new lines, is likely to produce an albino form wherever it is grown from seed in large quantities'. Nevertheless, in some species the odds against it would seem to be enormous. The Cretan *Crocus sieberi*, which has yellow-centred mauve flowers, took thirty years to produce 'Bowles' White', an immaculate and vigorous white form which was nevertheless an ample reward for such patience.

White in the wild

Contrary to expectations, species with white flowers and alba forms are not evenly distributed across the face of the earth. They are commonest in very cold regions – almost half the total species of flowering plants in the subarctic far north are white-flowered. In central Europe this drops to under a quarter, with even fewer in arid areas. They are also common in aquatic habitats. The reasons for this are not clear, though white flowers, especially when grouped together in flattish heads, are a favourite with flies, which may be more numerous than bees and butterflies in wetland and arctic conditions. However, it is difficult to generalise: white flowers occur in almost all families and range from specialised tropical orchids that depend on a single species of moth for pollination, to umbellifers such as cow parsley (*Anthriscus sylvestris*) which attracts all and sundry.

There are some quite localised areas where white flowers – or at least white-flowered forms – are particularly common. In parts of Hampshire, white-flowered plants of sweet violet (*Viola odorata*) are more common than the usual dark purple; and it has been observed that wood cranesbill (*Geranium sylvaticum*) has purplish-blue flowers in the mountains of southern Europe but the further north it grows, the commoner do pink and white flowers become, with some Scandinavian populations composed entirely of white-flowered plants.

Islands encourage genetic isolation, with unique forms and even different species from their mainland neighbours. On Crete the white form of *Anemone coronaria* appears to be commoner than the red, and the so-called white form of the dragon arum (*Dracunculus vulgaris*) can also be found (it is actually pale green instead of maroon). The Cretan endemic cyclamen, *C. creticum*, is pure white, and the island of Rhodes has a white-flowered variety (var. *rhodense*) of the normally dark pink *Cyclamen repandum*. No doubt there are many more examples; these are just ones I happen to have come across.

It is always interesting to discover albinos in the wild and to make observations on the frequency, time of flowering, insect visitors, etc. An article by Betty Molesworth Allen in the August 1986 volume of *The Garden* describes various white-flowered forms of Spanish plants. She records that white rosemary (*Rosmarinus officinalis*) has a longer flowering period than the blue and is always covered in bees, so they obviously do not discriminate against it. However, the white form of *Cistus albidus*, which is normally bright pink, also flowers longer but fails to set seed, suggesting that pollinators ignore it.

For the most part, white flowers are singularly visible. We see them clearly against foliage, dazzling in sunshine and luminous in shade. However, they are designed to attract the attention of pollinators, not of flower-fancying human beings, and some of these creatures cannot in fact see white as we do. Honey bees, for instance, are sensitive to ultra-violet light which in turn we cannot see. The ivory anthoxanthins in white flowers absorb ultra-violet light strongly and appear blue-green to bees. Many have veins or a radiating colour in the centre, which guide the bee to the nectar or pollen. They appear as a contrasting colour to us but show up to the bee because they reflect ultra-violet.

When bees are foraging, they work one particular kind of flower and no other until that source of food has been exhausted. Honey bees are particularly attracted to flowers which are blue to purple in colour, often with contrasting yellow to orange markings and a minty smell. The colour and/or scent act as signals to which the bee responds with single-minded devotion as long as those particular flowers are producing the food it needs. So what happens to a variable species such as Devil's bit scabious (*Succisa pratensis*) which, like the bluebell, usually has purplish-blue flowers, but occasionally produces white-flowered plants? Do bees remain constant to their favourite blue colour, or to all the flowers of that species in their collecting area, regardless of colour variation? Scientists observed a population of the scabious to find out how insects react to the odd white flowers. They discovered that individual honey bees varied in their response. Some visited only blue flowers one day and changed to white the next, but most worked both colours simultaneously. Bumble bees on the whole showed no clear preference. The responses of butterflies were quite different, however. They resolutely ignored the white flowers.

In another experiment, the visits of honey bees, bumble bees, syrphid flies and butterflies to wild radish (*Raphanus raphanistrum*) were recorded. This species again has variable flowers, with some populations predominantly yellow-flowered but others containing large numbers of white-

Lychnis chalcedonica – *the scarlet Maltese cross is pollinated by butterflies which, unlike most insects, can see red. The white form may well be ignored by butterflies in the wild and fail to reproduce. Many white forms owe their continuing existence to gardeners.*

behave in the same way as those observed in Britain, its 'alba' form is presumably ignored and remains unpollinated in the wild. Should this situation persist, it might happen that white-flowered plants would tend to die out. Alternatively, they might attract another pollinator altogether, which would eventually result in genetic differences and the evolution of a separate species.

White at night

Butterflies may find white flowers unappealing, but their relatives, the moths, generally prefer them. Indeed, many of the most highly scented white, cream and greenish-white or pinkish-white flowers are pollinated by moths: stephanotis (*Stephanotis floribunda*), tuberose (*Polianthes tuberosa*), tobacco flowers (*Nicotiana affinis, N. sylvestris*) and honey-suckle (*Lonicera periclymenum*), and all tropical orchids with white flowers and nectar spurs, to mention just a few. Some do appear to attract both butterflies and moths, however. The temperate pyramidal orchid (*Anacamptis pyramidalis*) and fragrant orchid (*Gymnadenia conopsea*) are scented and normally pale to deep pink, though both produce white flowers too. Both contain nectar in spurs which cannot be reached by short-tongued insects such as bees and flies. The colour attracts butterflies in the day and the scent, which is turned on more strongly at night, appeals to moths. Not all moths are nocturnal, however: the hummingbird hawkmoth (*Macroglossa stellatarum*) flies in the day and visits the same kinds of flowers as butterflies, such as pinks (*Dianthus* species).

The colour, scent, shape and position of flowers are all used by the plant in its advertising campaign to attract pollinators. Many flowers have a multi-purpose design which appeals to a variety of different insects, but some are more specialised and show adaptations to one group of insects, or even to a single species. Those which attract moths have a number of interesting characteristics.

Moth-pollinated flowers usually have a strong scent resembling cloves or perfumed soap. They are receptive to pollination at night, opening from twilight until dawn. Some close in the day or take on a shrivelled appearance, reviving as the light fades; others remain open but stop releasing scent during daylight hours. The tropical American *Cestrum nocturnum* has the local name *dama de noche* meaning 'lady of the night' (in other words, a prostitute) because it is unattractive in the day but seductively perfumed and beautiful at night. Nocturnal flowers are generally white or pale in colour and have fringed petals or deeply dissected lobes. These features help the moths locate them in darkness.

flowered plants. Again, butterflies strongly preferred the yellow flowers and ignored the white. In the case of these two species there is a variety of insect visitors so even if the white-flowered plants were ignored by butterflies, they would eventually be pollinated and pass on their genes. However, some species are much more dependent on butterflies. The Maltese Cross (*Lychnis chalcedonica*), a Mediterranean species, is one. Like many butterfly-pollinated plants it has red flowers, as butterflies – like birds, but unlike most insects – can see reds. If butterflies in its native Mediterranean regions

ABOVE Ipomoea bona-nox – *the moonflower opens at dusk and dies at dawn. It emits a sweet soapy perfume to attract moths. The scent is concentrated in bands (seen here as a star) to guide pollinators to the nectar.*

RIGHT Silene nutans *(Nottingham catchfly) – moth-pollinated flowers are commonly night-blooming. They open and release perfume from dusk to dawn, remaining closed and scentless during the day. Most are white or pale in colour, smell like scented soap and often have narrow or deeply divided petals. This nocturnal pink is found widely in Europe on dry calcareous ground.*

Scent is probably the initial attractant. It carries well in the still, humid night air and is detected over long distances. There is some evidence that the odours produced by night-flowering species resemble those put out by the insects themselves in order to attract a mate. It has also been suggested that the flowers actually look rather like moths to complete the illusion. More certain is the fact that moths can determine pale objects in the darkness and tend to fly toward sources of light. A dissected outline, as seen in the Nottingham catchfly (*Silene nutans*), might make the flowers more noticeable (and more moth-like), though more likely the shape guides the insect's proboscis to the centre of the flower as it begins to probe for nectar. Flowers which open at night do not have nectar guides in the form of contrasting colours or veins, though some species have highly scented bands of tissue radiating from the centre of the flower. These 'scent patterns' can be seen (or smelled, in the case of moths) as a star-like shape set into the circular white blooms of the aptly named moon flower (*Ipomoea bona-nox*).

Another characteristic of moth-pollinated flowers is that they are mostly held in a horizontal or pendant position and have no landing platform, as moths (unlike bees and butterflies) hover when feeding. They also produce more nectar than those visited by other insects as moths have a higher metabolic rate and need more food. The nectar is deeply situated in long hollow spurs or tubes which would be inaccessible to a bee but which the long proboscis of a moth has no difficulty in locating. In some species the tube is exceptionally long: that of the moon flower measures 3 in (8 cm) or more. Longest of all is the nectar spur of the Madagascan orchid *Angraecum sesquipedale*, commonly known as the comet orchid, which reaches 12 in (30 cm) – a remarkable length, though not quite living up to its species name *sesquipedale*, which means 'a foot and a half'. One of the best-known tales of natural history is how Charles Darwin and Alfred Wallace both saw this flower and concluded that it must be pollinated by a moth whose proboscis equalled the spur in length. At the time no known moth fitted this description but forty years later their prediction proved correct. A variety of an African sphinx moth was found with an extraordinarily long tongue and was named *Xanthopan morgani* forma *praedicta* accordingly.

Another famous story about a moth-pollinated species is that of the yucca. Several species of yucca are prized as garden plants for their bold rosettes of spiky leaves and spires of cream bells. In fact there are about forty different species, all native to America. The details of the story vary from species to species but the gist of it is that in the wild they are pollinated by moths (several species of *Tegeticula*) which spend their entire

Angraecum sesquipedale (comet orchid) and pollinator – Charles Darwin predicted that the pollinator of this giant Madagascan orchid, whose 12 in (30 cm) spur contains nectar only at its tip, must have a proboscis of equal length. Forty years later, such an insect was found: a tropical hawk moth which was accordingly named Xanthopan morgani praedicta.

lives in and around the yucca flowers.

The flowers are scented and in some species smell strongest at night. The female yucca moth homes in on a flower, climbs up the stamen and scrapes the pollen into a ball. She then lays her eggs in the ovary of a flower, after which she pushes the ball of pollen into the stigmas, which are joined into a tube. With the flower thus pollinated, the ovules begin to develop into seeds – apart from the ones containing the moth's eggs, that is. These develop abnormally and provide food for the growing larvae. As the normal seeds ripen, the larvae also mature and proceed to pupate underground. They emerge as moths during the flowering season, but not all at once. Yuccas do not flower annually, so the larvae of each brood hedge their bets and stagger their emergence over three years, so that the survival of the species is guaranteed even though one hatch may find no flowers. In spite of the fact that the moth parasitises the yucca's flowers, enough normal seeds develop to ensure that the yucca's – and the moth's – population is maintained. Their lives are completely interdependent.

Moths are not the only creatures to go a-pollinating at night. Bats are also on the wing and many in the tropics delight in floral feasts. They, too, seem to like pale flowers, especially white and creamy colours, though dingy shades of green and purple are also popular. In other respects they differ from moth-pollinated flowers in often being bell-shaped and relatively large and robust to admit the bat's nose and withstand its clutches. They also tend to hang free of branches and foliage to give clear access to large fast-fliers which hover when feeding. The bats are attracted to the flowers by rather sour musty smells (which are said to resemble those of the bats themselves), very sticky nectar and masses of pollen on protruding stamens that easily brush against the bat's fur. Few bat-pollinated plants are in cultivation, the majority being tropical trees such as the durian and the kapok. One of the few is the Mexican cup-and-saucer vine, *Cobaea scandens*, which is widely grown as an annual climber. Its bell-shaped flowers are usually greenish at first, then dull purple, but there is also an alba form with greenish-white flowers. They give out a sweet smell of fermentation. However, you will not be able to enjoy (or dread) the spectacle of bats among the flowers in your conservatory at twilight, as in temperate zones these interesting mammals feed on insects, not flowers!

Historic Alba

The beauty and symbolism of white flowers 'are such stuff as dreams are made on'. They also inspire action. Some of us are content to plant a garden and settle for enjoying white flowers and their associations in that way, but many have been brought to us by plant collectors and breeders who were actively interested – even obsessed – with the discovery of new white flowers. Others may get more out of creating images of them through painting, photography or other forms of art. If so, they are following in a long tradition, for white flowers have always had a special appeal to artists. Whatever the result – a poem, a painting, a garden, a photograph, a flower arrangement – it illustrates the power that flowers have over the imagination.

It was not until the nineteenth century, when botanical science and colonial exploration widened our horizons, that plants and flowers were regarded as interesting things in their own right. Before then, most of their importance lay in what they represented. Even in practical spheres such as medicine, their symbolic aspects were just as important as their physical properties. The world view has changed dramatically in the last century or two, leading us to value the rational evidence of our five senses above the imaginative and intuitive. As a result, many of the floral images in the art and literature of earlier times have little meaning for us, like familiar faces we cannot quite place. Discovering their identity and history not only adjusts our perceptions and judgements but also gives insight into the way other people and other cultures, past and present, have seen the world.

Heraldry and florigraphy

Though symbolic flowers often carry mystical connotations, they can also serve as thoroughly practical signs. Heraldry and florigraphy are very different in their techniques and purposes but both are methods of communication which make use of floral images.

Heraldry is a system of hereditary personal symbols denoting rank and allegiance. In western Europe it arose in the twelfth century, largely as a means of recognising feudal leaders and their retainers. A nobleman would adopt a symbol, known as a charge, which would appear on his seal and/or shield and be passed down to his descendants. The use of heraldic emblems became increasingly important during tournaments and battles when individual identity was obscured by armour. In the beginning, the symbols were mostly taken from the natural world and usually had some special meaning for the bearer. Easy recognition was the main factor in the design, so the plant or animal would be highly simplified and displayed in bright colours. Although straightforward in concept, the rules and terminology of heraldry became increasingly complex as arms were granted to more and more families and their generations of descendants, as well as to institutions and organisations, both secular and religious.

Albion and the white rose

In earliest times Britain was known as Albion, the white land. Some say it was named after the chalk cliffs of Dover; others because it abounded in wild roses. If the latter is the case, it seems particularly appropriate that the heraldic emblem of the English royal house is the rose. The device combines the red rose of the house of Lancaster with the white rose of York. The white rose is placed in the centre in acknowledgement of York's defeat in the Wars of the Roses, a dynastic struggle which dominated English history for thirty years of the fifteenth century. According to Shakespeare (in *Henry VI, Part 1*), the heraldic symbolism of the two factions originated in a quarrel between the contenders for the throne. The protagonists finally put their cards on the table and their supporters took sides by plucking either white roses or red roses from the Temple Garden in London. The Earl of Warwick, known as 'the Kingmaker' and a driving force in the struggle, together with Richard Plantagenet, Duke of York, picked white roses; and the Lancastrian Earls of Somerset and Suffolk, supporters of Henry VI, chose red. The two sides were finally united under Henry VII after the battle of Bosworth Field (1485). He was a Lancastrian of the Welsh family of Tudor, and strengthened the monarchy by marrying Elizabeth of York, the daughter of Edward IV who had been Henry VI's main rival. Whether or not there is any truth in the Shakespearean account (which was written a century after the events), the emblems are a vivid image of the struggle in which the 'pale and maiden blossom' was more than once dyed 'a bloody red' and have provided dramatic colour coding for productions of the plays which cover this period of British history. Historians maintain that the white heraldic rose, which is displayed as a *Rose en soleil* or *Rayed Rose* (i.e. a single white rose in the centre of a sun), was adopted by Edward IV after a Yorkist victory in 1461; and that

ABOVE Rosa x alba *'Maxima' (Jacobite rose) – a beautiful and resilient old rose, symbol of the Jacobites (supporters of the exiled Stuart king James II) in the 17th century.*

RIGHT Peristeria elata – *the Holy Ghost or dove orchid has fragrant waxy flowers whose lip and column together resemble a hovering dove. This Central American species is the national flower of Panama.*

the red rose emblem was created by Henry VII on his accession in 1485 as a means of stressing the final Lancastrian victory and the united future of the two houses. Thus the protracted conflict was only named the Wars of the Roses in retrospect. The peaceful solution was communicated not only by the new heraldic device, but also in his portrait which shows him holding a red and white rose. Subsequently a real red (well, pink) and white damask rose (*Rosa damascena* 'Versicolor') came on to the scene which became known as 'York and Lancaster'. Its flowers are sometimes pink, sometimes white and often bicolored. The story goes that it miraculously appeared in a Wiltshire monastery upon Henry's marriage to Elizabeth.

A white rose, *Rosa* x *alba* 'Maxima', is also the emblem of the Jacobites, who were named after the exiled Stuart king James II (the latin for James is *Jacobus*). They supported him and his descendants when claims to the throne were disputed during another troubled period of British history which resulted in the revolution of 1688 and continued well into the eighteenth century. The conflict caused factions throughout the realm. In Scotland the Highlanders wore white roses in support of 'Bonnie Prince Charlie' (Charles Edward) during the uprising of 1745 and their allegiance became immortalised in romantic songs which are still sung today.

Fleur-de-lis

The *fleur-de-lis* is familiar to many of us from childhood as the emblem of the Boy Scouts. The movement was founded in 1908 (by the British general, Robert Baden-Powell) and within a few years the *fleur-de-lis* badge, and accompanying motto 'Be Prepared', were known throughout the world. As a symbol of virtue, the *fleur-de-lis* has a long history.

There is some dispute over whether the *fleur-de-lis*, a highly stylised three-petalled emblem, is a lily, an iris or a dove descending. As far as heraldry is concerned, it is a lily. In addition to the tripartite emblem there is another which is known as a *lis-de-jardin*, in which the lily is depicted naturalistically. Although the symbol existed long before heraldry, it came into prominence when it was adopted as the French royal arms. According to legend, a lily was sent from heaven to Clovis, king of the Franks in the first century AD, on the occasion of his baptism. Clovis, an astute and ambitious monarch, put Paris on the map by making it the centre of his kingdom. The emblem is well documented from the reign of Louis VI, who came to the throne in 1108 and used it on his seal and coins. His son, Louis VII, took it as a symbol of righteous Christian purity when he went on the second

crusade to the Holy Land in 1147, though he was decisively beaten by the Islamic forces at Damascus. It has been suggested that *fleur-de-lis* is a contraction of, or pun on, *fleur-de-Louis*. Although the purported origin of the emblem was a white lily, it is depicted as gold in the case of the French royal arms, and red in the badge of the Italian city of Florence.

White flowers of nations, states and cities

Heraldry also has its more mundane side. Inn signs such as the 'Rose and Crown' and 'Rose and Thistle' originated in the banners which were hung outside when bearers of those arms lodged at the establishment in the course of their travels. The arms of institutions and organisations are emblazoned on buildings, vehicles, headed notepaper and other possessions, and most nations, states and cities boast a floral emblem, whether or not it is incorporated into their arms.

White flowers are often chosen as emblems because, in addition to their symbolism, white provides a strong contrast to the heraldic black, red, blue, green, purple and gold. Some are wild species, such as the British rose, Panama's dove orchid (*Peristeria elata*), Switzerland's edelweiss, Salvador's yucca and the Egyptian lotus. Others are white forms, perhaps because the species has a non-heraldic colouring such as pink, or possibly on account of their mystique. Examples of these include *Lycaste skinneri* 'Alba' of Guatemala and *Rosa virginiana* 'Alba' of the American state of Georgia.

Very occasionally, a horticultural variety is singled out for the honour. This happened in the Buckinghamshire borough of Slough whose armorial bearings include *Dianthus* 'Mrs Sinkins' – in heraldic jargon 'a swan proper holding in the beak a

The coat of arms of the Buckinghamshire borough of Slough – unusual armorial bearings which show a swan holding in its beak Dianthus *'Mrs Sinkins', a superlative white pink which was bred by a Mr Sinkins, master of Slough workhouse, in the 19th century.*

white pink slipped and leaved'. Mrs Sinkins was the wife of the master of Slough workhouse. In his spare time, Mr Sinkins raised new varieties of pinks. The heavily scented white one that he named after his wife became so popular that its contribution to civic pride was recognised by its elevation to the heraldic. Another example is the city of Exmouth which has *Magnolia grandiflora* on its arms, having some magnificent specimens of this species that is reliably hardy only in mild sheltered areas of the British Isles, such as Devon. A particularly fine variety of the species is named 'Exmouth' after the city which honours it.

The language of flowers

The Victorians really did say it with flowers. Today, roses still mean love but in the nineteenth century hundreds of different flowers had specific meanings. It was not just a sentimental Victorian pastime but the flowering, if you will forgive the pun, of a tradition that dates back to earliest times. Records of florigraphic signs exist from ancient Chinese, Assyrian, Egyptian and Indian cultures. The ancient Greeks used to send meaningful bouquets, some of their symbolism being derived from classical myths, such as that of Narcissus (see page 47 of Mystic Alba). There are Elizabethan verses which detail floral meanings – violets for humility, gillyflowers (i.e. pinks) for gentleness, as well as roses for love and lilies for chastity. The Victorian craze drew on these sources and on the writings of travellers such as Lady Mary Wortley Montagu who had visited Turkey and learnt the secrets of eastern flower language.

Florigraphy, which literally means 'writing with flowers', can be done with real flowers, illustrations or names. Further subtleties of communication were achieved by following rules such as leaving the foliage on the flower stem for the affirmative, and removing them to negate the meaning.

In 1840 the postal service as we know it was established. It led to a great increase in letter writing and the sending of greetings cards. The popularity of florigraphy grew accordingly, and book after book appearing to instruct budding florigraphers on the art of formulating and deciphering messages. After consulting the florigraphic dictionary, a rejected lover might send the following white-flowered note of despair:

Alas! I am CATCHFLY [betrayed]. I was impressed by your WHITE CHERRY [education] and THORNAPPLE [deceitful charms] but now I see only your NARCISSUS [egotism]. With STRAWBERRY FLOWERS [foresight] I should have realised the MEADOWSWEET [uselessness] of our MYRTLE [love]. I long for the LILY-OF-THE-VALLEY [return of happiness] but fear HEMLOCK [you will be the death of me].

Once a year, on February 14th, the florigraphic imagination really did run riot in a confection of Valentine cards. In 1850 the central post office in London at St Martin's-le-Grand had to cope with 70,000 such missives, many handmade and exceedingly intricate, with pop-up devices, lace, feathers, pressed flowers, painted decorations and, most of all, floral messages. Among the posies, sprigs and swags of the inevitable roses and forget-me-nots were fuchsias, symbolising taste, pansies [I think of you] and Indian jasmine [I attach myself to you]. But not all gushed with affection. Some were funny, rude or sarcastic, like the one which shows a lady looking up to a man-sized narcissus which has a man's head appearing from the topmost flower and the rhyme: 'You look all inexpressibly too, too. But such fearful utterness will not do.' Florigraphy could deflate egos as well as capture hearts.

The big white flower hunters

While Victorian and Edwardian drawing rooms witnessed precious sentiments about flowers, real romance was alive and well in the hands of the plant collectors, who combed hitherto unexplored regions of the world for botanical treasures. In 1869 the missionary Abbé David sent back specimens of *Davidia involucrata* which was later described in *Curtis's Botanical Magazine* as 'certainly one of the most striking of the novelties discovered in western China'. Also in China, George Forrest found a double white form of *Rosa banksiae* which was 'a hundred or more feet in length, thirty feet high and twenty through, a veritable cascade of the purest white backed by the most delicate green, and with a cushion of fragrance on every side. One such sight as that, and it is only one of many, is worth all the weariness and hardship of a journey from England.' Among his other white-flowered introductions were *Clematis armandii*, *Pieris forrestii* and *Jasminum polyanthum*.

One of the most extraordinary stories of this era is that of Benedict Roezl (1823–1885) who, though born in Prague, was employed for many years by the renowned orchid nursery of Sander in England. By all accounts, Benedict Roezl was a formidable man. He was keen on plants since he was a child but did not embark on the profession in which he was to excel until the age of forty-five – after he had lost an arm in a machine he had invented for cleaning hemp fibres. So, armed instead with a hook, which was to make him something of a legend on his travels, he set off for South America, and in spite of being robbed seventeen times and facing innumerable other hazards, he managed to cover a fair bit of it in his dogged search for new species.

The price of a priceless orchid

The story begins with another plant hunter, George Ure Skinner, who had discovered a magnificent orchid high in the tropical rain forest trees of Guatemala in 1836. It bore up to nine large flowers of a rosy-purple with a darker purple lip and was named *Cattleya skinneri* after him. At that point in time, no white cattleya had ever been found, but Benedict Roezl had a theory that all pink or purple cattleyas must, sooner or later, produce a white sport. He was undoubtedly right, though the rarity of such an event, coupled with the fact that these orchids mostly perch in the tops of exceedingly tall and tangled rain-forest trees, meant that his chances of finding one were incredibly small. But find one he did, and in a way that in itself is quite incredible.

In 1870 he was heading for the Guatemalan coast when a gang of unfriendly locals set upon him and dragged him off to their village where he caught a glimpse of something on the roof of the little mission church that looked remarkably like a white cattleya. Apparently the Indians collected unusual plants – exceptionally large-flowered, variegated, albinos, and such like – when out hunting and brought them back to grow on roofs and surrounding trees as ritual offerings, a practice going back to pre-Christian times. He therefore had good reason to hope that he had found the flower of his dreams.

Cattleya skinneri *'Alba'* – among the rarest of wild orchids, introduced from Guatemala in 1870 by the one-armed Benedict Roezl who acquired it from a cock-fighting padre.

He soon made the acquaintance of the *padre* and explained that his purpose in the area was to find a white *flor de San Sebastian*, as *Cattleya skinneri* was known locally. The *padre*'s reply confirmed his hopes: the white orchid on the church roof, among the host of oncidiums, masdevallias, lycastes and other cattleyas, was indeed *Cattleya skinneri* 'Alba'! It had been found by the Indians many years before and was the first and last they had ever come across.

Needless to say, it was more than his life was worth to desecrate the offering by stealing it, and beyond hope that the Indians would sell or give away such a prize. Instead, he walked off with the precious orchid as a reward for solving the village's most pressing problem: that of losing a major round in the inter-village cock-fighting tournament. Having trudged very nearly the length and breadth of the continent, Benedict Roezl was, if not an expert on cock-fighting, then at least one move ahead of the *padre* and his team. He told them he knew exactly why their cocks were losing and would advise them on what to do if they would swear to give him the white *flor de San Sebastian*. The situation was so desperate that they agreed to the bargain. He instructed them to stop shutting away the cocks in cupboards and instead to tie them up where they could see people – and hens. This they did with the result that the cocks became alert and aggressive, instead of confused and timorous, and went on to battle valiantly, while Benedict Roezl got away with one of the rarest orchids in the world. He sold it to a Mr George Hardy of Manchester for 280 guineas.

A royal lady and her price

An equally exciting story surrounds the discovery of the royal lily, *Lilium regale*, which was made in 1903 in China by Ernest Henry Wilson (1876–1930). Popularly known as 'Chinese Wilson', he made the most daunting explorations on behalf of Veitch's nursery in England and the Arnold Arboretum of Harvard College in the United States, introducing more new plants to western gardens than any collector before or since. Strangely enough, though he enthused about his new lily, American nurserymen showed little interest and its first introduction to cultivation was a failure. The botanists were equally unimpressed, identifying it as *Lilium myriophyllum*, rather than a new species – an error which took nine years to correct. In spite of this unpromising start, Wilson set off to China once again in 1910, determined to secure a massive consignment of bulbs for the horticultural market. The expedition travelled 1800 miles up the Yangtze (Chang) river, then a further 250 miles along the river Min into Tibet. The difficulties of travelling in the mountains were exacerbated by the fact that Wilson had to be conveyed by sedan chair, affecting the status of mandarin in order to gain respect of the local people and ensure some degree of safety.

Ironically, it was the sedan chair that nearly cost him his life. After digging up some 7000 bulbs, they began the return journey, which was made even more arduous by the precious cargo. To make matters worse, Wilson fell ill with dysentery and of necessity was confined to the sedan chair (he would normally get out and walk whenever the convoy was out of sight of villages and other travellers). On a narrow mountain track the expedition was hit by a rock avalanche which sent the sedan chair crashing down the mountainside a split second after he had clambered out of it. His quick escape undoubtedly saved his life but as he dived for cover, he was hit by a boulder which broke his right leg in two places. Still conscious but in severe pain, he then had to endure the entire mule train walking over him as the track was too narrow to allow them to turn round. Yet in spite of the terror of men and beasts, not one hoof or foot touched him as he lay immobilised in their path.

He managed to splint the leg with his camera tripod and was bumped along in another sedan chair for three days until they reached a Presbyterian mission. The doctor did what he could to set the bones but infection set in and he decided to amputate. Wilson refused and for three months battled against the odds for his leg to heal, finally realising that unless he returned to America, his chances of recovery were slim. He survived the long haul to Boston and went straight to hospital. The infection was treated successfully but his right leg was never the same again, being shorter than the left which gave him what he referred to as his 'lily limp'. It did not, however, prevent him from continuing his search for new species.

Lilium regale was the most outstanding – and memorable – of E. H. Wilson's 1000 or more introductions. Not one of the 7000 bulbs was harmed in the rock fall and the lily went on to become one of the most successful garden plants of all time. He paid dearly for his favourite but wrote that it 'was worth it and more … its beauty of blossom and richness of fragrance had won my heart.' Unfortunately, it did not soften his heart-lessness. Like so many plant hunters of the time, he suffered from bouts of colonial arrogance. In addition, the competition between them was so fierce that they would often strip an area of a certain species just to prevent a rival getting hold of the same novelty. Seen in this light, some of their introductions seem like the horticultural equivalent of the Elgin marbles. Wilson's response to the wild plant in its natural setting is not uncharacteristic:

> *'That such a rare jewel should have its home in so remote and arid a region of the world seemed like a joke on nature's part. However, there it was, and my business in life was to effect its transference to lands where its beauty would find proper recognition. Throughout the indefinite past, generations of the regal lily have lived unsung and unseen save by the rude peasants of a rude land. But few white men had passed that way when first I made my discovery and none had noted my royal lady. This had been preserved for me.'*

Lilium regale – *bringing back bulbs of the royal lily from China almost cost E. H. Wilson a leg, if not an arm. A rock fall fractured his right leg and, though he escaped amputation was left with what he referred to as his 'lily limp'.*

I wonder what those glorious flowers did mean to the local people and whether the bare bones of the mountains are still scented and shining with lilies once a year.

A free spirit

Compared with the introductions of E. H. Wilson, those of Reginald Farrer (1880–1920) are insignificant. What he brought back from wild and dangerous places was not so much botanical booty (though he was by no means a failure in this respect) as memories, impressions and feelings, which he has shared with us in his vivid and inimitable writings. His passion for plants was matched by a love of their habitations and a genial response to the human inhabitants of the region. In turn, he had an insatiable appetite for both visual detail and the words to describe it, inventing grammatical and verbal structures where conventional expression failed. Plants, both rare and commonplace, landscapes, weather, people, even domestic happenings were all grist to his adjectival mill; he took as much pleasure in describing 'a little aged woman going by us, with a wee white kitten enfolded in her grubby bosom', as in 'an enormous moon . . . coming palely up in a milky cloudwrack', or a viola's 'intensity of dark and velvety violence'.

Some of Farrer's most purple passages are lavished on white flowers, especially on white-flowered forms which stop even the most prosaic plant hunters in their tracks. If irresolutely unmoved by albino gentians which he found 'ugly and disappointing' (though he still searched for them), he is profoundly moved by other instances of these freaks of nature. On one occasion he came across some extremely rare white globe flowers growing beneath a smiling Buddha, 'of a perfectly pure white, rich and solid as cream or fresh ivory', which he named 'Trollius pumilus Perfectissimi'. At such a lovely sight he draws the inescapable conclusion that they have turned white through 'the influence of holiness'. His account of the harebell poppy (*Meconopsis quintuplinervia*) is another example. He explains that the normal form is lavender, with not one in a million plants showing any difference, though tending to pink or purple or blue ('rosy', 'amethystine' and 'azure' in Farrerese) in some populations. And among the 'flickering and dancing in millions upon millions of pale purple butterflies' and their 'innumerable soft blue laughter' there is suddenly a plant with white flowers, 'almost unimagineably beautiful, like living drops of purity or incarnate snowflakes – exactly what the Snowdrop ought to be, and isn't'.

The first albino harebell poppy Farrer found was dug up and taken back to the temporary garden he had made at his lodgings. Some time later he discovered another:

'I came on one as white and clean and clear as the soul of St John, a thing so beautiful as to give yet a new turn to the wheel of the day's delights. One tuft of a pure Albino we already had in the garden, though: so from this one, in reverent appreciation, I withheld the trowel, and left her swinging out her perfect snowy bells unperturbed. And I like to think of her still thriving through the years, and, still unperturbed, shaking out her snowy bells across all the din of the world, in a delicate and remote derision.'

The most singular example of a rarity that Farrer left to live out its life in the wild was the white tree peony. Tree peonies (*Paeonia suffruticosa*) had haunted plant collectors since they were first introduced by Sir Joseph Banks in 1787. There were reports that their natural habitat, the Moutan-shan in northern China, was flushed red with vast numbers, but subsequent introductions had all been garden varieties from China and Japan and collectors had found little or nothing left in the wild. For once, the depredations cannot be blamed on western plant collectors. Tree peonies are revered in the east as symbols of good fortune. They are grown for medicinal uses as well as for their beauty, and have been both cultivated and subjected to cult status for over a thousand years.

It was therefore in sheer disbelief that Reginald Farrer came across wild white tree peonies on a hillside in Kansu on the Tibetan border in 1914. At first he thought the white objects must be paper or wool, perhaps with some religious significance, as they looked far too big to be flowers, but as he plunged through the undergrowth and came nearer he realised that he was face to face with 'the most overpoweringly superb of hardy shrubs'. He must have realised how rare, valuable and in demand these plants were but typically brought back only a memorable description of 'the huge expanded goblets of Paeonia Moutan, refulgent as pure snow and fragrant as heavenly Roses', which at closer quarters revealed 'that single enormous blossom, waved and crimped into the boldest grace of line of absolute pure white with featherings of deepest maroon radiating at the base of the petals from the bosse of gold fluff at the flower heart'.

It was left to an American plant collector, Joseph Rock, to collect seed from a Kansu tree peony in 1925. He sent it back to the Arnold Arboretum and some years later seed of the original offspring were gratefully returned to him in China so that he could restock the wild peonies.

Apparently the poet Philip Larkin once said that happiness is white – meaning, I think, that he did not put pen to paper when he was on top of the world. Happiness was not just white to Reginald Farrer, but a 'blank white bliss' in the glory of the mountains and their flowers. Fortunately, it did not stop him writing!

White flowers in religious art

White flowers may induce ecstasy in plant hunters but in art and literature they express it, or its near equivalent. In religious art especially, the images of white flowers represent such states as joy, enlightenment, redemption, purity, innocence and divine love. The images tend to be stylised in the earliest examples – be it Buddhist, Hindu, Islamic or Christian. The artist was not concerned to represent one particular flower with any degree of accuracy, but distilled its characteristics into an idealised form which served as an easily recognised symbol. When few people could read, art was an important medium for teaching religious stories and practices, and the audience would derive meaning from details which hold little but aesthetic interest for us today.

Almond blossom, henna and lotus

For example, in a sixteenth-century Turkish miniature by Ahmet Nur b. Mustafa (which was copied from an earlier fourteenth-century version), the prophet Mohammed is depicted in prayer with Khadija, his wife, and his ten-year old cousin, Ali. Every pattern, colour and shape has a purpose: Mohammed and the boy wear green garments of revelation, lined with the black of mysticism; Khadija is dressed in a blue robe of purity with a red lining of love; and the walls and carpets are covered in intricate designs of interlocking stars which symbolise the knowledge of God in the soul (a constant theme in mosque ornamentation). As is the convention in Islamic religious art, the face of Mohammed is veiled in white. Not only is this a symbol of reverence but also recognition that representational art is inadequate when it comes to depicting what is beyond human imagination. (Indeed, Islamic precepts discourage representational art, hence the predominance of geometric patterns, stylised images and calligraphy.) The three figures have their hands folded – a gesture indicating submission to God in the Islamic prayer ritual. Behind the figures is an arcade of pillars through which are seen stylised trees. Two trees are in flower: one a peach, whose pink blooms signify the beginning of divine love (in contrast to the red of love which is shown in the lining of Khadija's robes); the other an almond, bearing white blossom which symbolises the truth of the Koran. And running downhill through the grove of trees is the river of mercy.

White almond blossom appears again in a very different painting of the late fifteenth century. It shows a group of whirling dervishes – devotees of the mystical Sufi sect whose dances induce ecstatic states in which they experience unity with God. They are shown in a flower-studded field, though in fact they would have performed the ritual in their convent. The outer circle of dervishes consists of musicians, observers and those who have dropped out of the dance, presumably from vertigo, and are being helped by others. The inner circle has just four dancers, each dressed in a different colour: brown, blue, white and red, which are the colours of earth, water, air and fire (the substances of material reality according to ancient

Greek philosophers). The topmost dancing figure is in blue and almost directly above him is a stem of four white lilies. Above the dervishes is an undulating skyline and a band of blue sky, against which grow almond trees covered in white blossom.

Lawsonia inermis (henna) – regarded by the prophet Mohammed as the flower of flowers for its scent and cosmetic properties.

Almond blossom may be a popular image but there is no single flower which occurs repeatedly in Islamic art. Mohammed himself apparently regarded henna (*Lawsonia inermis*) as the flower of flowers. Henna is a small spiny shrub native to the Middle East and eastern Asia, which reaches about 10 ft (3 m). It is widely cultivated in warm parts of the world and in greenhouses, being called mignonette tree in the United States, Jamaica mignonette in the West Indies (where it now grows wild) and Egyptian privet in England. Though best known for its leaves, which are powdered for use as a cosmetic dye, it also bears highly scented flowers which are distilled for perfume. In appearance they are not unlike those of privet, being small, four-petalled, creamy-white, and borne in panicles. It is said that Mohammed dyed his beard with henna, a practice which was handed down to the caliphs who succeeded him as rulers of the Islamic world. He was by no means the first to praise this plant. In the Bible it is referred to as camphire and King Solomon sings how his beloved 'is unto me as a cluster of camphire in the vineyards of Engedi' (a place on the Dead Sea coast where henna bushes still abound). And long before this it was valued by the ancient Egyptians, and on the Indian subcontinent as a dye, especially for fingernails.

Though important in many Islamic countries, henna does not appear to have much in the way of symbolic connotations or importance in art, other than in personal adornment. In contrast, the lotus is of great significance in both Hindu and

Paeonia suffruticosa *(tree peony) – 'the most overpoweringly superb of hardy shurbs', according to Reginald Farrer who rediscovered the wild species in Kansu in 1914, more than a century after its initial introduction by Sir Joseph Banks.*

Buddhist art. Indeed, the lotus is the central symbol of Buddhism (as described on page 46 of Mystic Alba), having been inherited from Hindu traditions and legends. Early Buddhist missionaries consciously used art to relay their message and familiarise people with details of the life of Buddha. Some of the finest examples are painted on rock shrines in Sri Lanka, which was converted to Buddhism in the third century BC. In many ways they are like giant strip cartoons with scene after scene illustrating the precepts and practices of the new religion. Gods and priests of the old Hindu religion, as well as beautiful women, are shown adoring the Buddha in reverential attitudes, which often involve the holding or offering of lotus flowers.

Renaissance roses and lilies of rebirth

In Christian art it is roses and lilies which are most used. They, in turn, were adopted from far older religions and traditions – from the Old Testament and from philosophies such as those of the Arabs, Greeks and Romans. Just who used what first is impossible to elucidate as boundaries and cultures merged over the centuries through conquests and trade. What is certain is that by the time of the Renaissance – that flowering of

European civilisation which began in Italy around the four-teenth century – they appear in almost every image of the Virgin Mary.

One of the major differences between the medieval world and the age of the Renaissance was in the approach to religious art and to images of the Virgin Mary in particular. In medieval times artists followed rules when making sacred images. The main purpose of art was to teach, not to express, and the authority and traditions of the church were absolute. As a result, the image was formal and the artists anonymous. Personal interpretation and realism did not come into it. Neither did symbolic flowers, to any great extent. And the iconography of the Virgin Mary was very limited, in keeping with the fact that mentions of her in the New Testament are few (nine in all) and reveal little about her.

Several different influences served to widen the medieval horizon. During the eleventh century in the south of France there began a revolution in thought, feeling and expression which became known as courtly love. The medium was the lyrical poetry of the troubadours: the message, a new style of romantic love which was to sweep through Europe and transform art and literature for centuries to come. It advocated a relationship between the lover and his lady in which the former's role was one of servitude while the latter was idealised beyond all criticism and reproach. In many ways, it resembled the role of vassal and lord in feudal society, a contract in which the vassal swore loyalty and paid homage in return for protection and livelihood. The lady was always superior and faultless – and always someone else's wife. (Romantic love and marriage were poles apart socially as well as ideologically, largely because marriage was then more to do with inheritance and property than feelings.) Infidelity was blasphemous and the whole affair was carried out in secret with much swooning, lovesickness and insomnia. The privilege of enduring this condition was more important than securing a return of affection, love being regarded as ennobling in itself. This belief ultimately led courtly love in mystical directions with the lover worshipping the god of love with the lady as a mediating saint.

The ideas behind this kind of romance were largely derived from the Arabic work *The Dove's Necklace* (*Tawq al-hamama*), a poetic treatise on love written by Ibn Hazm (994–1064) in Andalusia while Spain was under Moorish rule. In true Islamic mystical tradition, the work emphasised the spiritual side of passionate love.

It was not long before the wave of courtly love reached the shores of Celtic Arthurian legend. The two went well together, resulting in stories such as that of Lancelot, who was inspired by his (courtly) love of Guinevere to great chivalrous deeds. Meanwhile back in Italy, there arose a more devotional style of religious worship following the death of St Francis of Assisi in 1226. The scant biblical lives of Christ and the Virgin Mary were elaborated in various commentaries and meditations on the gospels and, in particular, the growing cult of Mariolatry was spread by mendicant orders of friars, such as the Franciscans. Not surprisingly, this movement was also influenced by the ideas, language and symbols of courtly love. At about this time too, artists began to emerge as individuals with a personal approach to religious subjects. Giotto (1266/7–1337) led the field in new interpretations but his influence was not fully felt until the fifteenth century when the ravages of the Black Death, which killed over 50 million people in the fourteenth century, had ceased its reign of terror.

A notable feature of courtly love literature is the imagery. It is full of flowers – cheeks like roses and lily-white brows – which became the standard fare of love poems for centuries. These lines from one of the Harley Lyrics (late thirteenth and early fourteenth centuries) are typical:

'A lily, lovely, lissom, slender,
Her white shot through with roses' splendour.'

And again:

'Her maiden-bloom's red, like the rose on the spray,
And lily-white loveliness shines in her face.'

The boundaries between secular and religious poems are blurred. The maidenly blushes and pristine virtue of the former occur with only slight modification in the latter. A nativity carol of the fourteenth century begins:

'Of a rose, a lovely rose,
Of a rose is all my song.

Listen nobles old and young
How this rose at outset sprung;
In all this world I know of none
I so desire as that fair rose.'

In this case, the rose is the Virgin Mary. She was known as the Rosa Mystica and the rose without a thorn (roses were traditionally without thorns until the fall of man in the Garden of Eden).

The association of roses with the Virgin Mary can be seen in many Renaissance paintings. The iconography is often com-plex. Not only the colours – red, white or blush – but the numbers of petals and rows of petals may have significance. The five-petalled single rose is an ancient symbol for the elements of earth, water, fire and air, together with ether, the medium which was believed to fill all space. It also has numerological significance: the indivisible numeral one being in antithesis to all other numerals; and the figure four including the perfect ten (one, two, three and four adding up to ten). The numeral one is taken to represent God, and four stands for the universe. Together they make the sign for one-in-all, the ultimate reality. The rows of petals indicated layers of reality, from the material to the spiritual, angelic and divine. Red roses were associated with the physical, often with Christ's suffering; white roses with the divine nature.

There are paintings of the Virgin completely surrounded by roses – the 'Virgin in the Rose Bower' by Stefan Lochner and the 'Madonna of the Rose Hedge' by Martin Schongauer (both fifteenth century) for example – and in Luca della Robbia's enamelled terracotta relief of the 'Madonna of the Roses', the infant Jesus reaches from his mother's knee to touch one of the roses, which all prominently display five pale petals – to us an endearing gesture though largely of symbolic meaning for the fifteenth-century observer. But more commonly the Virgin Mary is depicted with roses, either holding one or with vases of roses or with angels wearing roses. One of the loveliest is Giotto's 'Madonna and Child' which dates back to about 1320. The Virgin is holding a double white rose for which, again, the child is reaching. Among the strangest is the 'Assumption' by Andrea del Castagno, a painting completed about seven years before his death in 1457. In it a rather massive Virgin swathed in nun-like headgear rises to heaven, surrounded by a sunset mandorla, from a relatively small tomb which is filled with a *pot-pourri* of roses and flanked by a couple of nonchalant-looking saints.

Much of the appeal of paintings by Sandro Botticelli (*c.* 1445–1510) lie in his floral images. He is the master of Renaissance roses. Several of his Madonnas are pictured with roses but his best-known works are 'Primavera' and 'The Birth of Venus'. Though ostensibly classical, both of them are homages to a Venus who is Madonna-like. In 'Primavera' she is haloed by branches of myrtle (which, as described on page 46 of Mystic Alba, symbolised love and was dedicated to Venus) and raises her right hand in the traditional gesture of benediction, while at her feet are pink and white roses that fall from the dress of Flora, goddess of flowers. (Roses were also sacred to Venus, goddess of love and beauty.) In 'The Birth of Venus' the air is filled with floating blush roses, a motif repeated about five years later in the 'Coronation of the Virgin', in which they are showered into clouds of angels. If anything, there is more reverence conveyed through his 'pagan' Venus-Madonnas than in the pious coronation scene. His earlier works were in-

fluenced by the philosopher Marsilio Ficino, who attempted to unite classical philosophy and mythology with Christian doctrine in an educational ideal of *humanitas* – a system of values and practices which would result in the perfect human being. He believed that beauty is divine in origin and its contemplation leads to love, wisdom and enlightenment. Botticelli was commissioned to paint the mystical 'Primavera', in which Venus is synonymous with *humanitas*, for the Villa di Castello where the fifteen-year-old Lorenzo di Medici was educated according to these precepts. Botticelli's later works lost the exquisite beauty, languid grace and gentle, almost sexless sensuousness of these homages to the Venus-Madonna, becoming more intense and severe – and flowerless – as he came under the power of Savonarola, the puritanical reformer.

Lilies appear side by side with roses in many Renaissance paintings. In Raphael's 'Assumption and Coronation of the Virgin', which he painted when only about twenty (that is, around 1503), both lilies and roses are growing out of the Virgin's tomb, and Botticelli's 'Virgin and Child with the Two Saint Johns' shows the Virgin on an elaborately carved pedestal between tall vases of lilies and olive branches, whose bases are circled with red, white and pink roses. Just when this association of the Virgin Mary with *Lilium candidum* began is unclear. Once again, the origins may go back to classical myths (see page 49 of Mystic Alba). There is, however, a difference in the ways that roses and lilies are used, for roses are more for love, both secular and divine, whereas lilies almost always symbolise purity. More specifically, when they appear with the Madonna, they represent the dichotomy of virgin motherhood, the purification of the erotic through the mystic. Their shape suggests that of the vulva and their unvarying and immaculate whiteness signifies perpetual chastity and virginity. In addition, the flower opens into a six-pointed star. Stars of various shapes are potent symbols in almost all mystical traditions. The six-pointed star is most familiar as the Star of David or Magen David, but did not become the official seal of Judaism until the seventeenth century. It had a long history of symbolic associations, not least in the star of Bethlehem which guided the magi to the birth of Christ. The brightest star that dawned in the east was the planet Venus. In ancient times its rising signalled the coming of Astarte or Ishtar (also identified with Venus), the fertility goddess, who awoke Adonis, the vegetation god who spent half of the year underground. It was known as the Star of Salvation and regarded as a sign of renewal. Thus the lily in Christian art is a symbol of both the virgin birth and redemption or rebirth.

Not surprisingly, the lily is most used in paintings of the annunciation in which the Virgin Mary was first told of her role in the salvation of mankind. An early example is a wooden panel painted by Duccio around 1310 for the high altar of the cathedral in Siena. In formal Byzantine style the angel Gabriel extends his hand to the Virgin Mary above a two-handled vase which holds five stiff stems of lilies. About twenty years later, another Sienese artist, Simone Martini, painted an annunciation for the cathedral. Here the angel holds an olive branch and is crowned with olive, but on a level with his hands is an exquisite fluted vase of tall lilies with stylised spiky leaves and large widely spaced flowers. Though still predominantly gold, as was considered befitting for sacred images in medieval times, the graceful details and liveliness of this work were a great influence on subsequent painters of annunciations.

Almost every artist of the Renaissance produced an annunciation and virtually every one included the symbolic lily, whether in a vase or, as became increasingly popular, held by the angel Gabriel. Not all were paintings: the remarkable bronze doors of the Baptistry in Florence have their annunciation too, complete with lily, in one of the twenty quatrefoil relief panels. These were executed by Lorenzo Ghiberti who won a competition for the commission in 1401 and worked on the monumental task for twenty-one years. Neither were all Italian. Northern European masters are known for their remarkable interiors. The annunciation on the Mérode altarpiece by the fifteenth-century Flemish Master of Flémalle is no exception. It shows an extraordinarily domestic scene. The Virgin is reading a book, lolling against a carved bench, in a minutely detailed room with a wooden panelled ceiling, brass water pot, towel rail, open latticed window and a round table. Upon the table are a candlestick and a fancy jug which holds a very modest few-flowered stem of lilies, which looks as if it has just been picked from the garden. Were it not for the stoutly winged angel and the voluminous surreal folds of his and her garments, one could almost mistake this annunciation for a scene from everyday life.

No account of the Renaissance, however brief, can fail to mention the work of Leonardo da Vinci (1452–1519). His annunciation is memorable not for the Virgin, who is seated next to a remarkably ugly piece of carved furniture and surrounded by brickwork which is equally unattractive – in spite of the accuracy of the perspective – but for the angel who, with powerful feathered wings held horizontally, kneels on flower-studded grass as if he has just come to land. Behind him are the bold silhouetted shapes of a row of trees. His right hand is extended in a gesture of benediction and in his left is a stem of realistic lilies whose flowers are held almost touching his face. The ultimate Renaissance lily is still symbolic but painted with scientific accuracy. Leonardo made a number of botanical

A Lily by Leonardo da Vinci – the Eurasian Lilium candidum *is traditionally associated with the Virgin Mary and appears in many religious paintings. It became known as the Madonna lily in the 19th century to distinguish it from newcomers such as* L. longiflorum *and* L. regale, *oriental species which have proved much more adaptable in cultivation.*

drawings, one of which is a study of *Lilium candidum* which was marked for transfer to a canvas.

Though the lily is linked mainly with annunciations it does appear in other contexts. In Fra Filippo Lippi's 'Coronation of the Virgin' the crowded scene is set in a room beneath three massive arches. Long stems of lilies rise above the congregation, their white flowers framed by the side arches. The Virgin kneels under the central arch and on either side of her is a carved balustrade surmounted by a shell, as if to emphasise her womanhood, for which the shell is a symbol. Although the lily is principally the flower of the Virgin, it is occasionally seen in other company. St Joseph, husband of the Virgin Mary, is sometimes depicted holding the stem of lilies into which, according to tradition, his staff was turned. St Catherine is also shown with lilies, the story being that she converted her father, the Emperor Costis, by making the lily produce a wonderful perfume (it was supposed to be scentless before this miracle).

A Pre-Raphaelite flora

It may seem a jump from the Renaissance to the nineteenth-century Pre-Raphaelites but in terms of subject matter and floral symbolism they have a great deal in common. The Pre-Raphaelite Brotherhood was a group of artists who shared a common disgust with the way art had gone since the Renaissance, regarding Raphael as the start of the rot. They felt that his paintings were pompous, histrionic and devoid of real feeling. Their aim was to put the heart and soul – and nature – back into art, turning to poetry, religion, romance and 'real life' for inspiration and models. Among their favourite subjects were classical myths, Arthurian legends and scenes from rural life, the Bible and Shakespeare. Because of their themes and intentions, plants generally, and symbolic plants in particular, play an important part in their works. The Pre-Raphaelites took as great pains to get their plants botanically correct as to make their symbolism effective. Who can forget the lantern-lit weeds which have grown up against the door at which Christ knocks in Holman Hunt's 'The Light of the World'? Just as memorable are the wild flowers around the drowned 'Ophelia' which John Everett Millais completed in 1852. Among their numbers, both growing and strewn on the water are pale wild roses on the bank which tell of her trusting love; meadowsweet for useless-ness; daisies which stand for innocence (and may have first been used by Botticelli for this purpose in his 'Primavera'), as well as the inevitable forget-me-nots. All are painted with the same attention to detail that he bestowed upon the drowned Ophelia, for whom Millais used a partially submerged fully clothed model.

In addition to their way with plants – William Morris' wallpaper, Edward Burne-Jones' Merlin amidst the may blossom, Millais' children and autumn leaves – the Pre-Raphaelites are also associated with beautiful women. Typical are Millais' 'Bridesmaid', painted in 1851, and 'Beata Beatrix' by Dante Gabriel Rossetti (his name alone speaks volumes) which was

'Ecce Ancilla Domine' *[Behold the handmaid of the Lord] by Dante Gabriel Rossetti – a Pre-Raphaelite annunciation unusual both for its predominance of white and the way in which the angel Gabriel holds the lily stem first toward the Virgin.*

done over ten years later. Both have their eyes closed, shifting some of the emphasis to the floral images: orange blossom, symbolic of chastity and marriage, in the case of the brides-maid; and the pale opium poppy of death, brought by a haloed bird, for Beatrix.

Telling a story is another thing at which the Pre-Raphaelites excel. Arthur Hughes was taken with a poem by the Romantic poet Elizabeth Barrett Browning which tells the tale of Aurora Leigh, who aspires to become a poet and turns down a marriage proposal from wealthy cousin Romney because he has no interest in her poems. The painting shows her clutching the precious book of poems, Romney turning away, and a magnificent clump of white lilies, presumably symbolising her aspirations and purity of heart, in the left-hand corner. Another tantalising tale is told by Millais, this time derived from Tennyson, in which Mariana has been spurned by her fiancé after her dowry is lost in a shipwreck. In the painting, she stands up from her stool to stretch, rubbing her aching back and gazing out of the window to relieve the boredom. Autumn leaves lie scattered on the table and floor like her dashed hopes and in the stained glass window a snowdrop – emblem of hope and consolation – is depicted.

The religious paintings of the Pre-Raphaelites are likewise rich in symbolism. Two very different annunciations emerged from this era. The earliest, by Rossetti, is dated March 1850 and titled '*Ecce Ancilla Domine*' (behold the handmaid of the lord). As annunciations go, it is an extraordinary work and aroused considerable hostility. For a start, it is painted almost entirely in primary colours – red, yellow and blue – against a predominantly white ground. There is a window framing a blue sky and single slender tree, and a simple blue cloth screen behind the bed on which the Virgin Mary is crouched. The angel Gabriel takes up almost the whole of the left-hand side of the picture and is holding a lily, *stem first*, toward the Virgin whose eyes are fixed almost fearfully on the flowers. Above it hovers the white dove of the Holy Spirit. The gesture is most unusual: it is as if the angel is expecting the Virgin to take hold of the lily. Both figures are dressed in the plainest white sheet-like garments and have golden hair, very simple in style, and yellow haloes. Apart from the haloes they are essentially ordinary: a slightly frail teenage girl and a wingless, rather human angel. In the right-hand corner is an embroidery stand over which is draped a bright red cloth embroidered with a stem of lilies that hang upside down. Both this stem of lilies and the one held by the angel have three flowers. They are similar but different in that she is now expected to grasp the real lily by its stem, having previously embroidered its image. The only other bright red in the picture is a small but distinct touch – the virgin's lips. In both instances, the red is a reminder of the Virgin's humanity, that her actions are grounded in the physical though ordained and blessed by the divine. It contrasts markedly with the overwhelming whiteness and provides one of the few details in the painting – an unusual feature among Pre-Raphaelite works, which are characteristi-cally crammed with detail. As a result, it has a simplicity and directness that must have seemed little short of impertinent to conventional piety. Rossetti was particularly interested in symbols. He went as far as to compose a sonnet about his painting of 'The Girlhood of Mary Virgin' (1848–9) which explained that the pile of books on the floor represents virtues, on top of which is a tall stem of lilies in a vase, with three flowers in all, which symbolise innocence. Mary is sitting at her embroidery stand, perhaps working on the lily motif that was to appear, completed, in the annunciation painting which followed. She herself is described as 'An angel-water lily, that near God grows, and is quiet'.

Arthur Hughes' annunciation, painted eight years later, is very different. The figures are both standing, dressed in medieval-style garments and framed by a wooden arch inter-twined with vine tendrils (a reference to Christ's declaration at the Last Supper of being the true vine). The angel is more floating than standing, with hands crossed on breast and diaphanous wings extending down to the ground, midway in appearance between a cloak and a wedding dress. This strange device ends in lilies – an abundance of them – which are arranged around the feet and lower half of the angel. The effect is rather sentimental, the message of the symbol getting lost in profusion and prettiness.

Although the Pre-Raphaelites went to great lengths to show biblical scenes with a new and often homely realism, they were also inspired by the transcendental in nature and human life. Like the Romantic poets, whose generation overlapped that of the Pre-Raphaelites (Wordsworth died in 1850, the year Rossetti painted his annunciation), they found meaning and beauty in 'the meanest flower that blows'. Though usually shown in paintings of 'rural bliss', this belief is taken literally in Charles Allston Collins' 'Convent Thoughts'. In it a nun stands before an ornamental pool in a garden enclosed by a hedge-lined wall. In her right hand is a passionflower, a well-known symbol of Christ's suffering, at which she gazes thoughtfully. In her left hand is a book, held at her side with open illustrated pages that resemble a medieval manuscript. The implication is that her imagination and feelings are more stirred by the flower and its associations than by the elaborately beautiful book. The symbolism continues with the garden's white lilies, reminders of the Virgin Mary, and the pool's water lilies which signify the pure soul rising above physical limitations. There are even white goldfish in the water as if they too have been transfigured by grace. The homily extends to the frame, which is just as finely executed as the painting. Designed by Millais, it bears lilies in relief and the words '*Sicut lilium inter spinas*', a phrase from *The Song of Solomon 2*:

'As I am the rose of Sharon,
And the lily of the valleys.
As the lily among thorns,
So is my love among the daughters.

As the apple tree among the trees of the wood,
So is my beloved among the sons.'

This painting not only reveals the Pre-Raphaelite interest in the moral value of nature and the significance of symbolic images, but is also an excellent example of their painstaking attention to detail. Apparently each one of the six lily flowers took Collins a whole day to paint and nothing was left to chance – they and all the other flowers being painted from life in Oxford gardens. The sentiments may not have gone down too well but no lesser critic than John Ruskin commented that he had never seen the water plantain 'so thoroughly or so well drawn', though surprisingly – for Ruskin was a knowledgeable naturalist – he wrongly identified the plant in question. It is not water plantain (*Alisma plantago-aquatica*) but arrowhead (*Sagittaria sagittifolia*).

Stile floreale

The Pre-Raphaelite hankering to put the love back into loveliness waned, but the design possibilities of their images lived on in Art Nouveau, a movement which spanned the last decade of the nineteenth century and first decade of the twentieth. Beautiful plants – particularly lilies and roses – and beautiful women were swept into graceful sinuous lines, their three-dimensional reality and symbolic aspects subsumed into decorative compositions and delightful motifs that appeared in illustrations, stained glass, jewellery, wallpaper, textiles and furnishings. In Italy it was known as *stile floreale*.

In Art Nouveau the madonna lily flourished as never before: an illustration by Paul Berthon shows a red-haired maiden almost intertwined with stems of lilies, and a six-pointed star above her head; Eugène Grasset's design for the cover of the Christmas 1891 edition of *Harper's Bazaar* consists of a nativity scene heavily screened by tall stems of lilies through which little more is visible than the Virgin Mary's head and shoulders, the child Jesus in the cradle and two six-pointed stars. Eugène Grasset was one of the most versatile and influential exponents of Art Nouveau. In 1897 he published a book entitled *La Plante et ses Applications Ornementales* in which he expounded and illustrated his ideas of an art that he saw as 'a way of expressing our joy in living'. His great achievement was to grasp the spirit of the plant or flower without directly copying from nature or imitating past styles, though he acknowledged his debt to oriental, Islamic and medieval art.

White flowers in modern art

Although floral symbolism was most popular during pre-literate times and with movements which hark back to a pre-industrial golden age, it still plays an important, if less consistent, role in art and literature. Twentieth-century artists and designers exploit our subconscious associations with archetypal colours and flowers, as well as their enduring aesthetic appeal. The style and purpose may be new but the message has an ancient lineage: on a dark forbidding night, a bunch of arum lilies is placed on the seat of a car driven with panache to a mysterious destination; an immaculate arum lily is sprayed with different coloured paints. We are sold on romantic notions before we buy the car, and on the idea of natural purity as we choose additive-free cosmetics. Advertisers tease our responses but we are wooed more passionately by artists who dig deeper into meaning, require of us a longer attention span, and have no qualms about ambiguity.

The century was ushered in by Monet's water lilies whose reflections danced with light, and proceeded into the dark imaginings of Salvador Dali's myth of Narcissus. More recently, David Hockney's portrait of 'Mr and Mrs Clark and Percy' displays the cool enigmatic quality of white objects. The man's feet slide into the soft pile of the white rug while light plays on the detached beauty of the cat, a white balustrade and a vase of white lilies (in this case, a pure white form of the golden-rayed lily of Japan, *Lilium aūratum*) which, if not exactly symbolic, are definitely mood-inducing. The individuality of twentieth-century artists makes it impossible to generalise about the way that white flowers have been used in recent years. Yet in spite of the differences in approach, they show that that old white magic of sensuous body and spiritual soul, the erotic and the ethereal, is still casting its spell.

The arum lily: sinner and saint

In commercial terms, the white flower of the age is *Lilium longiflorum* but as far as art is concerned, it is the arum or calla lily (*Zantedeschia aethiopica*) – a flower deceptively simple in shape and more than usually complex in associations. It has been photographed in black and white by Ansel Adams, painted as part of a still life by Henri Matisse and given the Art Deco treatment by Tamara de Lempicka. Unconventional in shape, its sculptural qualities are however only one factor in its appeal. The central erect spadix is enfolded by a tubular, open-mouthed, skin-soft spathe, giving unequivocally sexual – even bisexual – connotations. Georgia O'Keefe's arum lilies are full frontal in their assault on our ideas of floral prettiness. One imagines that she painted them in much the same spirit as she

Art Nouveau madonna lilies – a nativity scene on the cover of the Christmas 1891 edition of Harper's Bazaar *by Eugène Grasset.*

gently amusing bisexual portrait of 'The Milliner' (Henri de Chatillon) who is trying on a lady's hat in the mirror. There are Indian flower sellers laden with armsful and bowed down with basketsful of opulent curvilinear blooms. He associates them particulary with the indigenous population of Mexico, so much so that they appear in his mural of 'The Great City of Tenochtitlán' even though, being South African, they would not have been known in Aztec times. A scene from the flower market also occurs in the background of one of his self-portraits. A small Indian girl with plaited hair and a spathe-like shawl kneels, arms outstretched, in front of a mountainous bunch of calla lilies. It is the same pose he used five years earlier in the crowning glory of his arum lily pictures: the 'Nude with Calla Lilies'. The model is kneeling, feet crossed, before an immense basket of flowers which fill the entire upper half of the painting. Her arms are extended, not to hold but to embrace. It is an unusually peaceful, sensuous work and one which is as much a portrait of the calla lily as a study of the nude model.

Arum lilies are associated with both weddings and funerals. An interesting example of their more spiritual aspects is in 'The Resurrection, Cookham' (1923–7) by Sir Stanley Spencer who lived in the village of Cookham, Berkshire, almost all his life. It is a huge mural-like painting over 15 ft (5 m) long and is remarkable for the way it combines the everyday with the imaginary. The church and its graveyard must have been a familiar sight to him, as to all the village's inhabitants. But for every commonplace object we have an image in our minds which may be changed, even transformed, by the associations and feelings it has for us. The graveyard in the painting shows how a simple view can become a complex vision and how both are integral and equally valid. Although the scene is filled with graves opening and the dead arising, there is nothing gruesome or awesome about it. Rather it is a relaxed, happy picture. The graves are mostly of creamy-white stone, like the church, which fills most of the upper background. Even the interiors of several graves are white and some of the dead are wearing white garments. But most of all there are white flowers – in bold swathes, dotted about, sprigged on the dress of a red-haired girl, in wreaths (this is where the arum lilies come in) and overarching the church door whose portals reveal three carved white *fleur-de-lis* (perhaps symbolic of the Holy Trinity?). You get the impression that the first thing the resurrected will do is to stretch and smell the flowers. There is something of everything here, from fields of daisies to formal florists' arrangements, but they are all used to convey one thing: joy – that mystical level of happiness which is so effectively communicated by the use of images such as white flowers.

posed naked for the photographer Seiglitz – though 'spirit' hardly seems the right word in such a physical context, for these magnificently solid and vivid but graceful flowers are a celebration of the flesh, the see-touch-smell-taste of earth-bound existence.

The Mexican artist Diego Rivera (1886–1957) painted calla lilies on a number of occasions. Many of his works are a striking blend of the sensual and intellectual, reflecting how deeply he came under the sway of both physical beauty and political ideas. Arum lilies feature largely in a number of his smaller paintings and even stray into his gargantuan murals, adding a touch of the opulent and exotic – and erotic – to portraits, details of everyday life and revolutionary scenes alike. He obviously enjoyed their simple sensual shapes which suited his bold direct style. He has them peeping from a corner of his

'Nude with Calla Lilies' by Diego Rivera – the shape of the calla or arum lily gives it inescapable sexual connotations.

*'The Resurrection, Cookham' by Sir Stanley Spencer
– a relaxed, happy picture of the dead arising amidst
swathes, wreaths and sprigs of white flowers.*

Mystic Alba

Life is dependent on light – on white light, which in itself is a rainbow of colours. From earliest times, human beings have sensed the importance of light and have regarded both light and the colour white as symbols of the highest good. White is the colour of beneficent power, love, purity, innocence and immortality. We attach particular significance to unusual or beautiful white animals and choose white garments and white flowers for many of our celebrations and rituals. White flowers, more than those of any other colour, have entered the numinous world of religion, myth and folklore.

A thirteenth-century Dominican monk by the name of Meister Eckhart wrote that 'God is not light nor life nor love nor nature nor spirit nor semblance nor anything we can put into words'. However true this assertion, it has not deterred human beings since time immemorial, including mystics such as Meister Eckhart, from attempting to describe the indescribable. The reason for such persistence in the struggle to express spiritual realities is that, instead of relying on unsatisfactory literal descriptions, we create deeply satisfying imaginative literature – myths, legends, fables, poetry – and in doing so manage to wrest meaning from the incomprehensible. Shedding light into the dark fear of the unknown, they somehow make the intolerable mystery of life more bearable – not as arguments to be proved or disproved, or statements to be believed, but as messages to be understood. We do not depend on the words themselves as much as on the images and associations conjured by the words – on reading between the lines. In this context, white flowers are not botanical or horticultural entities, but symbols to which we respond on largely subconscious levels.

Over the centuries these symbolic overtones have given many white flowers a certain mystique. They are, more than those of any other colour, flowers of power. A mystique is a powerful aura, something which can be suggested but which is beyond verbal and visual description. The words 'mystique', 'mystic' and 'mystery' all have their roots in the Greek *muo*, meaning to close the lips or eyes. In one sense this refers to the secret and inexpressible; in another to the contemplative act of withdrawing into the dark and silent inner space, a process which frees us from distractions and give insights into the intractable problems of existence. Originally mystics were those who had been initiated into divine knowledge – 'the mysteries' – and were sworn to secrecy. Nowadays we are fairly familiar with people or things which have mystique (even computer programmers have it), mysteries have lost their sublime element and become synonymous with puzzles, and mystics are often regarded with scepticism or derision. Nevertheless, mystical experiences, call them what you may, are still the source of much art, music and poetry, as well as religion and philosophy. They may even enter the gardener's world: Reginald Farrer, the great plant hunter, and Vita Sackville-West, gardener extraordinaire and poet, were mystics in their way.

We probably all have the capacity for mystical experience, for those timeless moments of ecstasy in which our world stands still and we feel part of everything around us. It may happen very briefly, or once in a lifetime, but leaves a profound impression. The everyday world of seemingly unconnected things and events suddenly appears as a skein of interrelationships and hidden beauty. The poet William Wordsworth, in his 'Lines composed a few miles above Tintern Abbey', called it a 'a serene and blessed mood' in which 'we see into the life of things'. Such moments are commonly inspired by natural phenomena, and on more than a few occasions, if symbolic literature is anything to go by, by white animals and white flowers.

Myths account for the origins of things in primeval time (known as 'dream time' to Australian aborigines), when the supernatural intervened to change the course of creation. They combine magic and realism in ways that both humanise the subjects and make them sacred. There is seldom any rational connection between the various subjects of a myth – human, plant, animal, the elemental or divine – but a psychological need is fulfilled by the beliefs and moral values they symbolise. The effect of a myth is to give a feeling of power or control, tempered by reverence, over what was previously remote from our lives. We befriend the outside world when it is made familiar through emotional significance: our bewilderment lessens as the cosmos takes on meaning.

In an age noted for its materialism, it is easy to be dismissive of mysticism, myth and the like, as things of the superstitious past. Inevitably many things have changed, but the 'real world' today is no different from that of our ancestors in that we can only grasp the nature of it through our senses and through the interpretation of experience. Each individual life is unique but is dependent for self-awareness and self-expression on the words and symbols shared with others, as much as on the DNA in the cells which is inherited. White lilies, roses or hawthorn are more than just flowers when seen through their significance to human beings for over three thousand years.

White flowers are immensely varied in their symbolism. At one extreme they represent death, and at the other, life. We will begin in this brief survey of them with sadness and end with gladness, though sometimes the boundaries are blurred, as in life.

White flowers of death and immortality

'And now he is dead and laid in his coffin
Six jolly sailor boys carry him on,
And six pretty maidens all carry white roses
So that no one might smell him as he passes along.'
(popular song: The Streets of Laredo)

Our attitudes to death and the dead are, to say the least, ambiguous. Death is the inevitable but unacceptable end of physical existence. When someone close to us dies, we are faced with conflicting emotions: between love for the person and a horror of the corpse; between perpetuating and breaking the ties. A death almost always requires major readjustments for the family and society, and grief may be accompanied by fear, guilt, anger or other feelings which are often largely repressed. The fact of death is simple, but the repercussions make it one of the most complex events in human society. Consequently, the beliefs, customs and rites concerned with death and burial are extremely diverse and tend to arouse deep feelings which we would rather avoid confronting.

The association of certain white flowers with death is just one example of our confusion. Though originally used as symbols of the soul liberated from physical limitations and entering eternal bliss (an aspect which makes funerals occasions of celebration in many societies), they often come to represent the mournful side of death as intense feelings of grief become identified with them. Scent can play a significant part in this transfer of meaning. Our sense of smell may be fairly feeble compared with that of other animals but it is sensitive enough to have a considerable influence in our lives nevertheless. Odours stimulate nerve impulses that are registered in the limbic system of the brain, an area concerned with memory and instinctive emotions. Smells tend therefore to be inextricably linked with past events and subconscious responses. The complexity of the sense of smell and its associations explains why the same scented white flower may be detested or adored by different people: feelings which may become enshrined in folklore as they are shared and gather momentum.

The belief in life after death has influenced responses to death and burial since prehistoric times. White, as a symbol of the spirit and immortality, has strong associations with death. In Judaism, the body is wrapped in a simple white shroud – which in popular folklore is what ghosts are supposed to wear. The early Christians regarded funerals as joyful celebrations and wore white, believing that the soul, being freed from the temporary confines of the body, would be united with Christ. In particular, the death of martyrs was a triumphal celebration which reaffirmed faith. By the eighth century the tradition had become muted and mourners wore black (as had been the custom in pre-Christian Greece). The use of flowers at funerals and shrines was initially condemned as a relic of paganism, but human weakness again prevailed and the tradition continued.

The traditional funeral flowers were commonly white because of their symbolism, but nowadays other colours are often chosen because white is considered 'morbid'. Likewise, there is a wide range of flowers which are chosen. Chrysanthemums are popular, so much so that in Italy they are almost exclusively used for funerals and are quite unacceptable as gifts. Highly scented white flowers were once especially favoured, partly for the practical purpose of disguising the odour of decomposition during the wake (a vigil over the dead person before burial). While awaiting cremation, the body of a Hindu may be covered with heavily scented mock orange blossom (*Philadelphus*) for this purpose. In societies where burial takes place very soon after death (as in Judaism), flowers, scented or otherwise, play little or no part in the rites. In British folklore, strongly scented white flowers with small petals that fall quickly are regarded as omens of death or disaster if brought into the house. Hawthorn (*Crataegus monogyna*) is especially feared for these reasons. Its rather sickly smell was described as a 'death stench' by the poet Sylvia Plath. Interestingly, the flowers contain trimethylamine, a substance formed when animal tissues begin to decompose. Though generally held as a symbol of hope (see page 42), hawthorn has a darker side. Its scent was said to remind survivors of the smell that enveloped the city of London during the Great Plague (1664–5) which killed 68,000 people (over a sixth of the city's population) in a year: a claim that ensured it a place in folklore as a flower of death.

White roses and lilies

Funeral customs vary greatly from region to region, even within the same religious framework. Until quite recent times, white roses had particular significance in funeral rites in parts of southern England. Washington Irving (1783–1859), the American essayist, recorded in *The Sketch Book of Geoffrey Crayon, Gent* that when a girl died, a wreath of white roses tied

Zantedeschia aethiopica (arum/calla lily) – *the whiteness and simplicity of the arum lily makes it a fitting symbol of spiritual values for both weddings and funerals.*

both the Greeks and Romans attached importance to it. Indeed, there are records not only of rose gardens being planted, but of the dying making a last request that the family should dine near the grave every year on the anniversary of the death and cover the tomb with roses from the bushes that had been planted around it.

Times change and few people today think of roses as being especially funereal. Their place has largely been taken by white lilies, whether true lilies such as *Lilium longiflorum* and *L. candidum* or the arum lily (*Zantedeschia aethiopica*). These are now the funeral flowers *par excellence*, especially for formal occasions. On the fortieth anniversary of V-E Day, 9 May 1985, the Soviet Union laid a wreath of arum lilies which was almost as tall as a man. Indeed, so funereal are arum lilies that in many countries they are becoming unpopular as garden plants. An Argentinian colleague told me that the superb, huge forms – such as Diego Rivera may have painted – have all but disappeared since his childhood! Outside the western world, lilies do not appear to have this significance. Apparently in the 1950s, a British military hospital in Malaya had to ask the local people to stop sending bouquets of *Lilium longiflorum* to the sick and wounded soldiers. Their distress at receiving what they thought of as funeral flowers must have been incomprehensible to the indigenous population. White flowers generally are not popular in hospitals and worse still are red and white flowers together. Known as 'blood and bandages', such a mixture is regarded as an omen of death. Military hospitals have even been known to issue written instructions forbidding the combination, in order to avoid a lowering of morale. (It can, however, have quite different connotations, being the colour of flowers used in Anglican churches at Whitsun to symbolise the fire and wind of the Holy Spirit.)

Iris albicans

The lovely white *Iris albicans* is the Moslem flower of the dead and is traditionally planted on graves. Originally Middle Eastern in distribution, it spread far and wide during the Ottoman Empire, a Moslem state which ruled much of North Africa, Asia Minor, the Balkans, southern Russia and parts of southern Europe during a reign which lasted from the thirteenth century until the end of the First World War. *Iris albicans* is a long-lived plant and can still be found around graveyards in regions that were once under Ottoman rule, including of course present-day Turkey. Gertrude Jekyll records in *A Gardener's Testament* (1937) that she brought back a plant from a Turkish cemetery on Rhodes, which is now a Greek island but was ruled by the Seljuk Turks from 1282–1309 and formed part of the Ottoman Empire from 1522–1912.

with white ribbons was carried before the coffin and after the burial was hung over her usual seat in church. 'They are emblematic of purity, and the crown of glory which she has received in heaven.' Sometimes black ribbons were used to signify the grief of the mourners and a white rose bush would be planted on her grave. The planting of roses on graves was an old tradition: John Evelyn, the seventeenth-century diarist, describes how young women would plant roses on the graves of their dead sweethearts; and William Camden wrote in his *Britannia* (1586) that this was an age-old custom even in Elizabethan times. Possibly it dated back to classical times, as

White flowers of love and hope

White flowers are as important for weddings as for funerals: an incongruity which is explained by their symbolism. They represent various aspects of the spiritual side of life: the soul, immortality, resurrection, joy, love, light, life, holiness, innocence, purity, virginity and so forth. Marriage in the modern western world may have more to do with love than kinship and property, but the traditions connected with it lag far behind and are still those inherited from ancient Greek, Roman, Hebrew and Christian practices.

In earliest times, marriage was governed by patriarchal laws. The Greeks and Romans gradually freed themselves from these restrictions and came up with a much more flexible arrangement by which a couple were considered married by virtue of living together as husband and wife – an arrangement which

Iris albicans – the Moslem flower of the dead, traditionally planted on graves. Originally Middle Eastern in distribution, it spread far and wide during the Ottoman Empire. Large numbers can still be found in Greek islands and other places where there are Moslem cemeteries. (Here seen in Turkey)

could be terminated as easily by either party. The rise of Christianity radically changed all this. In addition to the patriarchal Hebrew traditions it inherited, Christian marriage carried the burden of a high idealism. It was seen as a divine union, an insoluble bond in which fidelity and monogamy were absolute requirements. Marriage thus became more of a religious than a secular contract, with the church reinforcing or replacing the head of the family as patriarch.

The use of white flowers at weddings has had a similarly long and varied history, dating back to at least classical times. Long white dresses, bridal veils and bouquets of white flowers may seem little more than romantic notions today, but their origin lies in a practice which by today's standards sounds humiliatingly brutal. When status, kinship and property were the main purposes of marriage, proof of paternity assumed great importance and had to be demonstrated by the defloration of the virgin bride by the husband on the wedding night: an event proved the next morning when the bride publicly displayed the blood-stained sheet from her bridal bed. This element of Hebrew tradition eventually became ritualised in certain motifs of the Christian wedding: the bride enters the church veiled and wearing garments which are essentially sheet-like; at the end of the ceremony the bride is unveiled and with a triumphant smile throws away her bouquet. The white flowers symbolise her virginity and its loss is appropriately known as 'deflowering'.

Fortunately, white flowers stand for more than virginity and still predominate in weddings for their associations of love and joy, even when all-white is avoided because of its now almost obsolete associations with virginity. At present the most popular white flowers include gardenias, freesias, stephanotis, lily-of-the-valley, lilies, carnations and roses, all of which are beautifully scented. As symbols of hope, light and life, white flowers of various kinds also play an important role in a number of different festivals and celebrations.

Hawthorn

'Mark the fair blooming of the hawthorn tree;
Who, finely clothed in a robe of white,
Feeds full the wanton eye with May's delight.'

(Geoffrey Chaucer)

Hawthorn, whitethorn or thorn bushes (*Crataegus monogyna*) go under a number of common names, including quickset (from their use as a 'quick' or living hedge), bread-and-cheese (from the tasty new leaves which, like all country children, I remember eating each spring), and may (from the time of flowering). Its profusion of white blossoms weigh down the hedgerow branches, making it a symbol of joyous hope throughout most of Europe and a feature of countless May Day celebrations.

The custom of going 'a-Maying' dates back to pre-Christian times and was at its height during medieval and Elizabethan England. On the first of May everyone would be up at dawn to gather branches of may blossom which would be borne in triumph and merriment back to the towns and villages for decorating houses, doorways and churches. It was a day out in the spring countryside observed by all ages and classes, from the poorest peasant to the royal family and courtiers, when everyone, according to John Stow, 'would walk into the sweet meadows and green woods, there to rejoice their spirits with the beauty and savour of sweet flowers and with the harmony of birds praising God'. At the end of the day, even the grimiest urban street would be fresh with blossom and greenery.

The celebrations involved choosing a May queen (sometimes a May king too, who in some areas was known as the Green Man, or Jack-in-the-green), crowning the sovereign with may blossom, and setting up a maypole, from the top of which streamed different coloured ribbons. Dancing round the maypole was the highlight of the day, each person holding a ribbon and weaving in and out of the other dancers to criss-cross the ribbons. When the dance ended, the ribbons were plaited in a colourful pattern all down the pole. It is delightful to watch, but is quite a complicated routine for the dancers. In large towns and cities, maypoles were permanent fixtures. The last one in London, which stood near Somerset House in the Strand, was removed in 1717.

May Day customs have their origin in ancient religions which celebrated the coming of spring with various rites to ensure fertility in the year ahead. They are associated with Beltane, the Celtic May Day festival when people jumped over, and drove cattle between, ceremonial fires, and with the Roman goddess Maia who seems to have had something to do with increase and new spring growth, and whose name could be the origin of our word 'May'. Over the centuries, the traditions were modified and tended to lose their meaning. The ritual of crowning a king and queen of the May, for instance, may once have been a much more serious business in which they were honoured through the growing season, then put to death at the end of the year. (The need for sacrifice is common to many ancient fertility rituals.) The ambiguity of hawthorn as a symbol of hope and omen of death (see page 39) could have its origin in a racial memory of this practice. Perhaps the strange ceremony of 'bawming the thorn' is connected with it too. ('Bawming' is a dialect word for 'anointing'.) In July an old thorn tree would be decorated with red and white flowers and ribbons, then the village children would dance round it. One of the last villages to hold this event was Appleton Thorn in Cheshire.

Another fertility rite involving hawthorn has been recorded in Herefordshire. On New Year's Day the women would weave branches into a globe which was then hung in the farm kitchen for a whole year to ensure good crops. At dawn on the following New Year's Day it would be removed by the men, taken outside into a field (usually the first one to be sown) and burnt in a rite which was called 'burning the bush'. While the men did this, the women were indoors making a new globe for the forthcoming year.

'Bringing home the May' by Henry Peach Robinson – the custom of going 'a-Maying' dates back to pre-Christian times.

The Glastonbury thorn, which flowers at Christmas, is world-famous. The Somerset market town of Glastonbury is identified with King Arthur's Avalon, the Holy Grail and many other mysteries. According to legend, it was in Glastonbury that St Joseph of Arimathea founded the Christian church in England. He reached Wyrral (Wearyall) Hill, Glastonbury, on Christmas Eve and there stuck his staff into the ground. Immediately it rooted and began to grow, and ever since has miraculously bloomed at Christmas. This phenomenon was first mentioned in a poem dated 1502. In 1753, when the Gregorian calendar was introduced, the new Christmas Day fell twelve days earlier. Large crowds gathered to see whether the thorn would take note of the change and when it failed to do so, blooming as usual on the old Julian calendar Christmas Day (now 5 January), many refused to go to church.

The Glastonbury thorn has been named *Crataegus monogyna* var. *praecox* and is thought to be a sport (mutation) of the species. There are records of early-flowering thorns in other parts of the country (at Quainton in Buckinghamshire and at Orcop in Herefordshire, for example) which probably came from cuttings of the original Glastonbury thorn. Attempts at propagating it from seed have, however, failed: they all flower in May.

Snowdrops

*'Lone flowers, hemmed in with snows and white as they
But hardier far. . . .'*

(William Wordsworth: *To a Snowdrop*)

The pristine white snowdrop (*Galanthus* sp.) flowers in winter. Its fragile appearance belies an exceptional hardiness, with leaves and flower buds tipped with tough points to pierce the frozen earth and delicate petals resistant to severe frosts. According to legend, it was created by an angel. Adam and Eve were driven from Paradise, so the story goes, in a blinding snowstorm. They had never experienced the harsh flowerless world of winter before and were so terrified at the cold bare landscape that God took pity on them and sent an angel to assure them that growth would begin again when spring returned. The angel took a handful of snowflakes and threw it to the ground where they turned into snowdrops which, with their tracery of green on the inner petals and hint of fragrance, are reminders that the freshness of spring is not far off.

In spite of their role as harbingers of spring (symbols of hope and 'a friend in need' in Victorian florigraphy), snowdrops are also regarded as unlucky if brought into the house, because of their shroud-like shape and whiteness and the fact that they often grow in graveyards. A more charitable view is that they are dedicated to the Virgin Mary as they start to flower on 2 February, the feast of the purification of the Virgin and presentation of the child Jesus in the temple. This feast is also known as Candlemas Day because it is the day on which church candles are blessed. It used to be celebrated with a procession of girls dressed in white who sang: 'The snowdrop, in purest white array, First rears its head on Candlemas Day'.

Galanthus nivalis – *snowdrops are symbolic of hope and according to legend were created by an angel to reassure Adam and Eve that spring would return.*

Orange blossom

Some traditional wedding flowers have gone out of fashion or perhaps are unavailable from florists. Orange blossom has, for example, long been used in wedding bouquets and hair ornaments, both for the sweet fragrance of its waxy white flowers and their association with innocence and chastity. It flowers all year round and is used a great deal in countries where citruses can be grown outdoors (they will not tolerate frosts). In the nineteenth century, the custom spread to northern Europe, encouraged no doubt by the vogue for orangeries (large conservatories) where citruses were grown

for their fruits as well as for the enjoyment of the incomparable fragrance of their blossom. Perhaps we can now look forward to a revival of the tradition, now that both conservatories and citrus trees – albeit on a more modest scale – are popular again.

Jasmine

'The jasmine throwing wide her elegant sweets,
The deep dark green of whose unvarnished leaf
Makes more conspicuous and illumines more
The bright profusion of her scattered stars'

(William Cowper)

The scent of jasmine (jessamine) flowers is one of the loveliest in existence. A number of different species of *Jasminum* have fragrant flowers, one of which (*J. officinale*) is reasonably hardy in northern Europe. However, there is no comparison between the perfume of this jasmine flowering outdoors in a cool British summer and that of the tender species (such as *J. sambac* or *J. polyanthum*) in the warmth and humidity of a tropical evening. In seventeenth- and eighteenth-century Canton (now Kwangchow or Guangzhou) the jasmine gardens stretched for miles along the banks of the Chu Chiang (Pearl River). Before dawn, armies of women would begin to pick the buds which had lost their pink flush and were therefore about to open. Then numerous workers employed by the flower merchants would string the flowers into hair ornaments, lamp shades, garlands for buildings and other fragrant articles. The flowers and their delirious fragrance would last all day and through the night, the perfume intensifying with the warmth of the body and the humid evening air. Every summer the city was 'like snow in the night, and was fragrant everywhere'. Jasmine is symbolic of amiability and friendship. It was presented to guests and made up into scented balls for those with hangovers.

There is perhaps more of seduction than innocence in its use as a hair ornament which remains fragrant and fresh till dawn, but its whiteness and perfume, if nothing else, make it eminently suitable as a wedding flower. It is used as such in Asia and is also popular in parts of Italy – a tradition that folklore attributes to the Grand Duke of Tuscany, who in 1699 obtained a remarkably fine double-flowered form of what was presumably *J. sambac* from Goa (a coastal region of western India that was once part of Portuguese India). The flowers were so rare and exquisitely scented that he prized it above anything else and refused to give away cuttings or allow its propagation. Needless to say, his gardener could not resist the temptation

ABOVE *Orange blossom is highly scented and has always been popular in wedding bouquets.*

and stole a piece for his fiancée on her birthday. This enterprising young lady fully appreciated the value of the gift and carefully tended the precious cutting. It did so well that she soon had enough to start a profitable business. The proceeds enabled the couple to marry (which on a gardener's wages had previously seemed a distant possibility) and to this day Tuscan brides 'as delicate and fair as the jasmine flowers they wear' remember the traditional saying that 'she who is worthy to wear jasmine is as good as a fortune to her husband'.

Myrtle

Myrtle (*Myrtus communis*) is one of those plants that seem familiar enough in name but which few people can recognise. It grows wild in western Asia and Mediterranean regions and is

RIGHT Myrtus communis – *myrtle was dedicated to the goddess Venus who was often worshipped under the Roman name for the plant,* Myrtilla.

easily identified by its neat aromatic evergreen leaves and most delicate scented white flowers whose five petals are almost hidden by a mass of stamens. In former times its beauty and fragrance made it of major importance in the Middle Eastern and countries bordering the Mediterranean.

The Romans dedicated myrtle to the goddess Venus, who on occasions was worshipped under the Latin name for the plant, *Myrtilla*. She was supposedly wearing a crown of myrtle when she arose from the sea (see page 51). Conquering heroes who achieved bloodless victories were permitted to add myrtle to their crowns of bay (*Laurus nobilis*). Its wood is exceedingly hard and was once used to make weapons, in contrast to its foliage and fragile flowers which symbolise peace. According to Pliny the Elder, the Romans and Sabines – an ancient people who went down in history for the 'rape' of their women by Romulus's men – laid down their arms beneath a myrtle tree.

The North African Moors who conquered Spain in the eighth century brought with them the Middle Eastern love of myrtle. The Alhambra, their ancient palace and fortress, has a Court of the Myrtles (*Patio de los Arrayanes*) and the surrounding parkland was planted with orange trees, myrtles and roses.

Myrtle is mentioned many times in the Old Testament. Again it symbolises peace. The angel of the Lord appeared to Zechariah in a myrtle grove to foretell of the new Jerusalem, and the prophet Isaiah promises that the penitent 'shall go out with joy, and be led forth in peace . . . and instead of the brier shall come up the myrtle tree'. It has remained significant for Jewish people and is used ceremonially (along with the citrus, palm and willow) in the Feast of the Tabernacles (Sukkoth) as a thanksgiving for the fruitfulness of the land. According to one writer in Victorian England, myrtle was in such demand by Jewish customers that some nurserymen grew it exclusively for them. The Hebrew name for myrtle is *hadas*, meaning 'sweetness', from which we get the girl's name Esther.

Myrtle also symbolises love. Its use in wedding bouquets may have come from ancient Greece and Rome or from the Middle East. There is a tradition that the sprigs of myrtles in the bouquet should be planted in the newly weds' garden and if they root then the marriage will be a success. Lady Diana Spencer's bouquet of gardenias, stephanotis, lily-of-the-valley, freesias, *Odontoglossum* orchids and Mountbatten roses (all white but for the golden roses), also included sprigs of myrtle taken from bushes which had been grown from Queen Victoria's bouquet.

Like orange blossom, myrtle is no longer readily available but should increase in popularity as more people come to know it as a trouble-free, attractive conservatory plant with a romantic history.

The lotus

> '*And the pool was filled with water out of sunlight,*
> *And the lotus rose, quietly, quietly,*
> *The surface glittered out of heart of light . . .*'
>
> (T. S. Eliot)

From earliest times the lotus (or lotos) – a kind of water lily – has been regarded as a symbol of fertility, life and immortality or rebirth. Water lilies are among the loveliest flowers in the world and the fact that such translucent beauty arises from mud suggests the duality of body-and-soul upon which the complex symbolism of the lotus in Hinduism and Buddhism is founded.

The word 'lotus' does in fact refer to several different species, principally to *Nymphaea lotus*, a night-flowering species with pure white petals, which is known as the white or Egyptian lotus and was sacred in ancient Egypt; and *Nelumbo nucifera* (formerly *Nelumbium nelumbo* or *Nelumbium speciosum*) which has circular peltate leaves and white to pink flowers held high above the water on spiny stalks, followed by flat-topped 'pepper pot' seed capsules. The latter is the sacred lotus of India, China and Tibet – sometimes confusingly called the Egyptian bean (because of its edible seeds). *Nelumbo* is the Sinhalese word for the lotus. Both the sacred lotus and the Egyptian lotus are highly scented.

Nymphaea lotus is the national flower of Egypt and since earliest times has been revered as a symbol of the Nile, giver of fertility and life. In ancient Egypt it was sacred to Isis, the goddess of nature whose cult spread to Greece and Rome in the third century BC. In a matrilineal society, she was the most important of all deities and wife of Osiris, god of death and the underworld. He was killed by his brother Set, god of evil, but Isis brought him back to life, symbolising the annual cycle of growth and death (or dormancy) in the natural world. The ancient Egyptians believed that the soul lived on after death and mortals could be reborn from lotus flowers as gods. Egyptian lotus flowers were cut for decorating banquet tables, and multi-necked vases were specially designed to hold the heavy flowers upright and separate.

In ancient Egypt the lotus was also associated with Horus, son of Isis and Osiris, who was the sky god. He is usually represented as a falcon whose eyes are the sun and the moon; sometimes as a winged sun. The lotus flower closes at night and re-opens at dawn. This, together with its ray-like petals, suggests similarities with the sun. The rising sun was personified by Nerfertem, a god who rose from the lotus every morning and withdrew into it every evening.

The sacred lotus (*Nelumbo nucifera*) is venerated throughout southeast Asia. A Hindu myth tells how it was created from

In Buddhist mythology, Buddha himself first appeared floating on a huge lotus and is traditionally depicted on a lotus throne.

the supreme being's navel, floating away on the flood which destroyed the world (myths of overwhelming floods are common to many cultures). Inside the flower sat Brahma, the god of creation and integration, who turned the lotus into a new world, its petals forming mountains, valleys and rivers. Brahma is the first god of the Hindu trinity, the others being Vishnu, the maintainer, and Shiva, the destroyer – or, as we might better understand it today, the recycler. Together the three deities are worshipped by repetition of the sacred syllable *Om* and by the phrase *Om mani padme hum*, which means 'Hail to the jewel in the lotus'. *Padma* is the Hindi word for lotus. It is also associated with Lakshmi, wife of Vishnu and goddess of prosperity and beauty. Veneration of the lotus extends into secular life: the *Padma Shri* is the highest award in India for distinguished service.

In Buddhist mythology, Buddha himself first appeared floating on a huge lotus. It is figured as a throne – the unusual flat-topped capsule suggesting a seat and representing the centre of the cosmos. Buddha is frequently depicted sitting cross-legged on the lotus throne (which is often highly stylised) with the soles of the feet resting on the thighs: a position known in yoga as the 'lotus position'. The lotus germinates in the darkness of mud and water and rises above the surface to unfold in the light: a process analogous to the growth of consciousness. It therefore represents purity and enlightenment, and in a wider sense is a symbol of the sun and its life-giving light. In Tantric Buddhism, however, the emphasis is sexual, with the lotus standing for female energy which unites with the male principle (represented by a thunderbolt) in procreation. Its Tantric associations are echoed in everyday usage: in Japan the flowers and stalks are eaten to give fertility; and in traditional Chinese medicine the fruits and seeds are taken for problems of the reproductive organs and in child-

birth. (Lotus-eaters are not, however, those who eat the sacred lotus, but people in Homer's *Odyssey* who ate the fruits of the lotus tree – possibly *Ziziphus lotus*, a North African species – to induce indolent forgetfulness. Nowadays the term refers to those who spend life in idle pleasure or in a dream world.)

Narcissus

'And Narcissi, the fairest of them all,
Who gaze on their eyes in the stream's recess,
Till they die at their own dear loveliness.'

(Percy Bysshe Shelley)

The Greek myth of Narcissus and Echo is a more than usually sad one. Narcissus was the exceptionally handsome son of Cephissus, the river god, and the nymph Leiriope. Tiresias, the blind prophet of Thebes, foretold that he would live long unless he once looked upon his own face. (In ancient times there was a widespread belief that it was unlucky or even fatal to see your own reflection.) A nymph named Echo fell in love with him. She had become embroiled in one of Zeus's affairs, being persuaded by the powerful god to keep his wife Hera chatting so that she could not spy on him. When the deception was discovered, Echo's chattering was suitably punished: from then on the only thing she could utter was the last syllable of someone else's words. Worse still, Narcissus spurned her love, and she pined away until nothing remained but her empty voice.

In turn, Narcissus was punished by the gods for his cruel rejection of Echo. He was condemned to fall in love with his own reflection and, unable to tear himself away from gazing at his own face in a pool, he too slowly died, fulfilling the prophecy that he would not survive seeing his own reflection. After his death, a lovely narcissus flower appeared where he had languished away. The mythological flower has given us the psychoanalytical term 'narcissism', meaning an obsessive interest in one's own body or an erotic self-admiration.

In contrast, the narcissus is the Chinese New Year flower and a symbol of good fortune and success. The species used for this important celebration is a variety of the 'Paper White' narcissus, *Narcissus tazetta* var. *orientalis*. Though not native to China, it was recorded as early as the tenth century when it was probably introduced by Arab traders from western Asia. Vast numbers are produced in the Fukien area of southern China, specially treated for flowering in the home at New Year. They are sold as clusters to grow on pebbles standing in water, sometimes in kit form with instructions for 'carving' what is known as a 'crab's claw narcissus'. This is done by carefully cutting away a large part of the fleshy onion-like scale on one side of the bulb, then placing it horizontally over the water, cut side uppermost, so that the flowers emerge sideways and on dwarf stalks. The flowers are pure white with a yellow centre and have a delightful scent.

In Chinese folklore, all flowers are feminine, as are the colours white, yellows and soft blues. Trees, together with shades of red, pink and purple are masculine. Each season has its own symbolic flower: the narcissus for New Year; white-blossomed plum for winter; peonies for spring; the lotus for summer; and chrysanthemums (or osmanthus) for autumn.

Osmanthus

Though little known in the western world, the oriental genus *Osmanthus* includes some attractive slow-growing shrubs with small evergreen leaves and tubular scented flowers. They are related to the olive and are sometimes referred to as 'fragrant olive' or 'holly-leaved olive'. *Osmanthus fragrans* is especially popular in China where the fragrant white flowers are used to scent tea. It blooms in the summer but is associated with the moon festival and autumn. There is a legend that the pattern on the moon's surface is an osmanthus tree. Many moons ago, one of the immortals, Wu Kang, was banished to the moon as a punishment for infidelity and told to chop the tree down. However, the tree is also immortal and can heal itself instantly. So Wu Kang chops forever and the tree remains as lovely as ever, and on bright moonlit nights the seeds of the osmanthus fall to earth so that we too can enjoy its beauty.

Camellia

Tea is probably the single most popular drink in the world. It is produced, in all its varieties, from the leaves of *Camellia sinensis*, an evergreen shrub grown throughout tropical and subtropical Asia and parts of Africa and South America. It bears attractive five-petalled white flowers with a boss of yellow stamens and looks like a rather small version of the familiar garden camellia. The origin of tea is given as follows: the Indian monk Bodhidharma, founder of the Zen school of Buddhism, came to China in AD 520 and spent nine whole years in silent meditation at the Shao-lin Temple on Mount Wu-t'ai. Once during this time he fell asleep. When he awoke he felt so ashamed at this lapse that he cut off his own eyelids and threw them to the ground. Where they landed the first tea bushes began to grow. Their leaves are indeed the shape of eyelids and the caffeine they contain certainly helps one to stay awake.

The garden camellia has a much larger flower than the tea bush. A number of species are grown as ornamentals but commonest is *C. japonica* and its varieties and hybrids. It is a variable species whose scentless flowers may be red, pink or white. In Victorian florigraphy red and white camellias were regarded respectively as emblems of the somewhat abstruse qualities of 'unpretending excellence' and 'perfected loveliness'. Camellias were at their height of popularity in Victorian times. They adorned clothes, hair and hats, featured in embroidery and jewellery, and starred in Alexandre Dumas's play *La Dame aux Camélias* which he adapted from his first novel in 1852 (and which later became the theme of Verdi's *La Traviata*). It was centred around his mistress, Marie Duplessis (Marguérite Gautier in the play), a French courtesan. The title was inspired by the fact that she only ever wore camellias because she could not stand scented flowers (they made her cough). Enigmatically, she wore white camellias for twenty-five days of each month and red for the remaining five, presumably as a personal symbol of her monthly cycle.

White lilies

'To gild refined gold, to paint the lily,
To throw perfume on the violet
Is wasteful and ridiculous excess'
(Shakespeare: *King John IV*)

There are a number of different species of lily which are naturally white, and others which have white forms, but the most important by far in terms of mythical and religious associations is *Lilium candidum*, popularly known now as the Madonna or Annunciation lily. It did not, however, get these names until the nineteenth century when other white-flowered species (such as *Lilium regale*) were introduced and it became necessary to distinguish between them. *Lilium candidum* is found wild in the Balkans and Asian Turkey (ancient Asia Minor). At one time it was such a common garden plant in the British Isles that John Gerard, the Elizabethan herbalist, referred to it as 'our English white lilly'. English it certainly is not, but its cultivation spread to northern Europe at a very early date, having previously been introduced to countries bordering the western Mediterranean from Asia by the Phoenicians, who dominated trade in the region during the first millennium BC. There are records of the plant from 2350 BC and its image occurs on Minoan, Assyrian and Egyptian antiquities. Its perfect beauty and whiteness were undoubtedly what made it so important then, as now, but it was also valued as a medicinal plant for skin complaints and as a perfume ingredient.

The Romans regarded *Lilium candidum* as a symbol of modesty, majesty and dignity. They dedicated it to Juno, queen of the Olympian heaven, who was also goddess of women, brides and childbirth, and the wife of Jupiter. The origin of the lily was given in a myth which tells how Jupiter wanted to make

Narcissus poeticus – *a Greek myth tells how this flower arose beside a pool where the handsome youth Narcissus pined away with love of his own reflection.*

the hero Hercules immortal. To do this, he gave Juno a sleeping potion and while she slept he put the infant Hercules to her breast so that he would drink the milk of immortality. When the baby finished feeding, her milk continued to flow. The drops which fell to the earth sprang up as white lilies and those which ran into the heavens became the Milky Way (*Via Lactea*).

Many pagan myths and celebrations became absorbed into Christianity. It is possible that the Roman dedication of *Lilium candidum* to Juno, queen of heaven and goddess of all that is essentially female, became transferred to the Virgin Mary, who is also known as queen of heaven. In its turn, Christianity created an origin myth for the lily. It describes how the disciple Thomas was not present when the Blessed Virgin died. He arrived on the scene after the burial, so he asked the other disciples to open the tomb in order that he might pay his last

respects. They did so, only to find that her body had disappeared and in its place white lilies were growing.

The Virgin Mary is especially revered in Roman Catholic communities, and wherever she is venerated, Madonna lilies are likewise favoured. Traditionally they are used to decorate shrines and churches on her feast days: notably 25 December, the birth of Christ; 25 March, the Annunciation; and 2 July, the festival of the Visitation, which celebrates the visit of the Virgin Mary to her cousin Elizabeth, mother of John the Baptist. There are numerous examples of the close association of Madonna lilies with the Virgin Mary, from the coat of arms of the Borough of Marylebone (Mary-the-Good) in London to the religious order called the Order of the Blessed Lady of the Lily, which was founded by Garcias, fourth King of Navarre (the Basque kingdom of southwest Europe, now divided between Spain and France). Apparently, he was miraculously cured of a serious illness after he had a vision of the Virgin Mary, who appeared to him in a flower of the Madonna lily.

And from Seville in southern Spain comes the legend of the simple-minded boy. He was the only son of a poor widow and a happy, lovable child but quite unable to learn anything at school. In despair, his mother finally appealed to the abbot of the local monastery to take him as a lay brother. He worked devotedly at his manual tasks but his education made little progress. All he could learn was to join his hands in prayer and repeat, 'I believe in God; I hope for God; I love God'. Every evening, when he finished work outdoors, he would retreat to the chapel, go down on his knees and endlessly repeat these three phrases in absolute contentment. Then one day he failed to appear in the chapel. The monks searched and found him dead in his cell, lying on the bed of straw, his hands clasped in prayer and a blissful expression on his face. He was buried in the cemetery with a marble cross for the tombstone, engraved with the words 'I believe in God; I hope for God; I love God'. Not long after, they were mystified to find a lily growing on the grave. In curiosity the monks opened the grave and found the roots of the lily coming from the heart of the boy.

According to Christian cosmology the white lily symbolises purity, virginity, innocence and benediction, all of which are exemplified by the Virgin Mary in the annunciation, conception and birth of Christ, and in her virgin motherhood. It also signifies resurrection and life after death. The early Christian communities of the Byzantine Empire and Asia Minor lived where this striking plant grows wild. To them, its cycle of subterranean dormancy, renewed growth and resplendent flowering came to symbolise the life of the soul, as the lotus did for Buddhists. It also became associated with Easter and the death and resurrection of Christ. (In northern Europe, *Lilium candidum* has now been superseded by *Lilium longiflorum* as the Easter lily because the former does not flower in time outdoors, whereas the latter can be grown under glass to flower in early spring.)

Lilies are commanding plants. They have a definite presence – it could almost be described as a soul – which ensures that they stand out like a peal of bells, whether in the wild or in cultivation. They are the most sumptuously elegant of wild flowers and their bold and graceful shapes, large richly scented flowers, exquisite colouring and perfect waxy texture have scarcely been improved upon by hybridisation. Not surprisingly, the lily also symbolises perfection.

White roses

'Frail the white rose and frail are
Her hands that gave
Whose soul is sere and paler
Than time's wan wave.

Rose frail and fair – yet frailest
A wonder wild
In gentle eyes thou veilest. My blueveined child.
(James Joyce: *A Flower Given To My Daughter*)

In contrast, few roses in cultivation are wild species. Their perfection is a triumph of the plant breeder's art, stretching back through so many centuries that it is seldom possible to trace their ancestry with any certainty (though recent advances in genetics are beginning to unravel some family histories). Roses may be synonymous with fair English beauties and inseparable from images of thatched cottages and rural bliss, but like Gerard's 'English white lilly', their origins are also in the Near East.

In Islamic culture, the rose reigns supreme as queen of flowers, the epitome of perfection. Roses occur repeatedly in Islamic poetry. One legend (in the *Book of the Nightingale* by the poet Attar) tells how the nightingale's exquisite song is inspired by its ecstatic love of the rose, and how all roses were originally white until the nightingale, in its passion, flew repeatedly into a bush of white roses until its blood dyed them crimson.

The image of paradise as a garden is common to many cultures. One fourteenth-century Turkish poet, Yunus Emre, has its inhabitants – streams, birds, trees and flowers – chanting the name of god. The rose does it by wafting its scent. It is the scent which is particularly significant in Islamic culture. The origin myth tells how roses came into being from the sweat of the prophet Mohammed, presumably because it was as sweet as perfume. A nineteenth-century Turkish writer described how in a dream Mohammed smelt of saffron and roses and his two sons, Hüseyn and Hasan, of white roses and carnations respectively. The association of sweet scent with holiness is found in various cultures. Indeed, there are records of several Christian saints whose bodies gave off a pleasant fragrance long after death. It was known as the 'odour of sanctity'; an expression which has passed into general usage as a derogatory term for piousness.

So important were roses in the Ottoman Empire that the Topkapi palace gardens received an order for 50,000 white roses in 1593, and the even larger gardens at Edirne were required to supply 16 tons. Not surprisingly, Turkish sultans were invariably portrayed smelling a rose and they may very well have smelt of roses too – though probably not for reasons of piety. Turkey was long the main producer of essential oil of roses, known as attar or otto of roses after the Persian word *atir*, meaning perfume. Oil production was centred on an area of the Ottoman Empire which was later to become Bulgaria, where rose oil is still an important industry.

Roses and their scent were used not only as luxuries by the ruling classes, but were also regarded as sacred. Fallen rose petals were never allowed to lie on the ground because of their association with Mohammed. Rose water, together with camphor, is sprinkled on the dead before burial. It is also used as holy water to consecrate mosques. Saladin, the sultan of Egypt and Syria, and adversary of Richard I of England in the Third Crusade, captured Jerusalem in 1187 and had the mosque of the temple – which the Christians had been using as a church – washed throughout with rose water before he would enter it. Likewise, Mohammed (Mehmet) II, Ottoman sultan of Turkey, captured Constantinople (present-day Istanbul) in 1453, marking the end of the Byzantine Empire, and had the church of Hagia Sophia cleansed with rose water to convert it into a mosque.

Roses are also an important part of everyday life in many Muslim countries. Turkey continues its devotion to roses with rose-petal jam, Turkish delight flavoured with rose water, and the delightful custom of sprinkling a visitor's hands with rose water – an everyday ritual performed by bus conductors, shop keepers and waiters for those they serve. Romance and roses are inseparable. In Turkey, the giving of pink roses (or carnations) shows 'I like you'; red ones mean love; and white indicates a proposal of marriage. I was once rather taken aback by being given red and white together by a Turkish admirer. Making a joke about the seriousness of the message, I was relieved to be told that the florist had run out of pink and had suggested red and white together as a suitable substitute!

Islam reached Spain in the eighth century, Greece, Yugoslavia and Bulgaria in the fourteenth century, and with it came the roses. *Rosa damascena*, the damask rose or rose of Damascus, with pink-budded white flowers, is Asian in origin. But before this, the Roman invasions had introduced roses to western and northern Europe in the first and second centuries AD. The Romans had a great passion for roses too and used them in enormous quantities, growing them in heated beds during the winter for an all-year-round supply. They wore them in garlands and strewed petals on wine and on the floor during banquets, supposedly to prevent intoxication. On one occasion, this was taken to such excess that some guests actually smothered under the floral confetti.

In contrast to the connotations of pleasure, the rose was used in ancient times as an emblem of secrecy. Harpocrates, god of silence (a later version of the Egyptian falcon god Horus the Child), is portrayed holding a white rose, with one finger on his lips. The present-day expression 'sub rosa' or 'under the rose' means 'in secret' and this has its origin in the ritual of hanging a rose over the doorway or table to denote that everything said on that occasion would be strictly confidential. Rose-shaped ornamentation, especially around a central ceiling light, is derived from this symbolism and is still referred to as a rose.

In Roman mythology, roses had their origin in Venus, the goddess of beauty and love. There are several versions of the myth, which describes her love for the handsome Adonis (who originally was a Syrian vegetation god). In one, she treads on a thorn on her way to meet him, and her blood turns into red roses. In another, Adonis is killed by a wild boar, and his blood becomes red roses while white roses spring from Venus's tears. Yet another tells how she treads on a thorn when rushing to help the wounded Adonis and her blood dyes the white roses red. Whichever version you prefer, they all associate roses with both love and pain.

Venus is identified with the Greek goddess Aphrodite who in turn has links with Ishtar, the great mother, who was the Babylonian and Assyrian goddess of love and fertility. We may think of Venus/Aphrodite as the supreme embodiment of feminine beauty, but the many myths surrounding this powerful deity suggest a more primeval sensuality. According to the Greek origin of the universe, the offspring of Gaia (earth) and Uranus (sky) were imprisoned within the earth by their tyrannous father. She groaned in agony so that eventually one of them, Cronus, with his mother's knowledge, rebelled and castrated his father. He threw the genitals into the sea which foamed around them. From the foam arose Aphrodite, goddess of natural beauty, sexual love and procreation: a scene romanticised by the fifteenth-century Florentine painter Botticelli, in his 'Birth of Venus'. She is borne on a white shell, symbolising feminity, and around her float blush-white roses which signify love. Earlier interpretations were more powerful than pretty: she was depicted with a beard and worshipped in transvestite rites and ritual prostitution which celebrated the mystic unity of male and female in the act of procreation. Lucretius, the Roman poet and philosopher, called her the 'sole ruler of the world' through whom we come 'to the shores of light, to joy and love'.

The son of Venus/Aphrodite was Cupid/Eros, god of love (in the sense of directed affection), who is represented as a naked winged boy with a bow and arrow. As we are reminded every

Valentine's Day, he shoots through the heart anyone he condemns to the agony and ecstasy of infatuation. The images – and physical presence – of roses are part and parcel of the scene, symbolising love and joy, tinged with suffering, as they have since earliest times.

Inevitably the rose found favour in Christian imagery. A rosary (from the Latin word for a rose garden) is a series of repeated prayers – usually ten *Aves* ('Hail Marys') beginning with a *Paternoster* ('Our Father') – counted on a string of fifty-five beads. The Virgin Mary is the Rosa Mystica, symbolised by a white rose, and the red rose represents the blood of Christ, with its five petals denoting his wounds. Gerard Manley Hopkins wrote a poem entitled '*Rosa Mystica*' which explores this image:

> '*What was the colour of that Blossom bright?*
> *White to begin with, immaculate white.*
> *But what a wild flush on the flakes of it stood,*
> *When the Rose ran in crimsonings down the Cross-wood.*'

Rosa damascena is often associated with the Virgin. The flowers of this Asian species are pink in bud, fading to white, a colouring which provides a perfect image for body and soul. One variety of the damask rose is called 'Leda', recalling the myth of Leda, queen of Sparta. Zeus, the ancient Greeks' god of nature (Jupiter to the Romans), fell in love with this mortal woman and came to her in the form of a swan. Whoever named this damask rose grasped the significance of its colouring: its dark pink buds representing human flesh and blood; its open white petals symbolising the swan, in its turn an incarnation of the god of life. In the case of the Virgin Mary, the pink and white are emblems of her physical conception and birth of the divine Christ and occur frequently in illustrations of her. In the vision of St Bernadette at Lourdes, the Virgin Mary's feet were each adorned with a damask rose.

In Christian legend, the origin of roses is given in a story about an innocent girl who was wrongly condemned to be burnt at the stake. As she entered the flames, they were immediately quenched: the burning branches became red roses and those not yet alight were transformed into white roses. There is also another story of how white and red roses came about: the Virgin Mary laid her veil on a rose bush to dry and the red roses beneath it were changed to white.

White roses and angels

There is a French legend that moss roses were created by an angel. The moss rose, *Rosa centifolia* 'Muscosa' is a mutation of *R. centifolia* (the cabbage or Provence rose). These roses have moss-like growths covering the sepals which, at a glance, can look disconcertingly like aphids but are in fact oil glands that give added fragrance. Moss roses come in a variety of colours from deepest crimson to flesh pink and, of course, white. The double pure white *R. centifolia* 'Alba Muscosa' (the White Bath or White Moss rose) appeared in 1790 and the ivory 'Blanche Moreau', which has dark brown moss and thorns, followed in 1880. Mutations such as these appeared as if by magic – or, of course, by angelic intervention. The white Cherokee rose, *Rosa laevigata*, which is the state flower of Georgia, was the work of the *nunnshi*, the 'little folk' of North American Indian mythology. Though not exactly angels, they were on the side of the angels when they turned a lovely young girl who had been captured by a hostile tribe into a white rose, complete with defensive prickles, to keep her from harm.

There are many flowers which are called roses because they resemble them. The Christmas rose (*Helleborus niger*) is one. This member of the buttercup family (Ranunculaceae) has long been a garden favourite as it flowers in winter. Its leathery dark evergreen leaves and waxy five-petalled white flowers are remarkably resistant to frost, though they may suffer from marking in cold wet weather. The flowers do indeed look like roses, with five petals and a centre of yellow stamens, but are unscented. A German legend describes how one Christmas Eve, a girl was weeping in the forest. She had been sent out to pick flowers for the church (white, of course, to celebrate the birth of Christ) but could find none in the snow-covered countryside. Taking pity on her distress, the Archangel Gabriel appeared and told her to look under the carpet of dead leaves. She did so and found Christmas roses.

A far more complex use of rose imagery is in Dante's *Paradise*, the third book of *The Divine Comedy*. It is the climax of his journey through hell, purgatory and heaven, guided in the last instance by his beloved Beatrice. Dante is a visionary – and highly visual – poet. He attempts (and succeeds perhaps better than anyone else) to put into words the essentially inexpressible. Light, life, love, nature, spirit and ultimately God, are the very subjects of *The Divine Comedy*. His great work is shot through with images of light which culminate in his description of the heavenly host of angels and saints in the 31st canto. The vision is in the form of a vast multi-petalled white rose of unimaginable whiteness whose centre is a flight of angels with golden wings and faces of living flame which cast no shadows in the divine light. In common with Meister Eckhart, Wordsworth and all others who have had similar mystical experiences, Dante sees 'the scattered leaves of the universe' fused together in a single light, an insight that happens in a timeless moment and fills him with peace and contentment. To express the inexpressible, he chooses a white rose.

'Leda' – a damask rose with deep pink buds and
white full-blown flowers, aptly named after the Greek
legend in which Zeus fell in love with a mortal
woman and came to her in the form of a swan.

White Flowers in the Garden

'I love colour, and rejoice in it, but white is lovely to me forever'
(Victoria Sackville-West)

Colour in the garden can be incidental – something which evolves over the years from chance purchases and casual sowings – or it can be developed with the same consideration given to interior decor or personal appearance – or even a work of art. Gertrude Jekyll (1843–1932) called it 'thoughtful care and definite attention', 'a state of mind that will not tolerate bad or careless combination or any misuse of plants'. This may sound extreme but basically it is something we all recognise, perhaps not consciously, when we admire the successful designs and plantsmanship of gardens open to the public. Such achievements often display, but do not necessarily require, considerable amounts of land, time or money: Miss Jekyll's principles can be applied to a flat suburban rectangle just as well as to a rolling country estate, and in many ways the smaller and less well-endowed a garden the greater the challenge.

Most gardeners just grow what they like and like what they grow, in many cases preferring a riot of colour above all else. Bright colours are exciting, especially in countries where clothing and decor (and weather) tend to be sombre, and to some extent the intrinsic beauty of the plants covers a multitude of aesthetic sins. But Miss Jekyll would have had little patience with this approach. She, and notable gardeners before and after her, regarded gardening as an art form – hence the lofty aspirations – and depended on a highly developed sense of colour, bright or otherwise, and appreciation of form, as well as on green fingers. 'The aim,' she wrote, 'is always to use the plants to the best of one's means and intelligence so as to form pictures of living beauty.' She grew what she liked but learnt by careful observation what best to do with it: 'It has taken me half a lifetime merely to find out what is best worth doing, and a good slice out of another half to puzzle out the ways of doing it.'

Gardening at this level is more than time-consuming: it can become all-consuming. There is no point at which one can pronounce it finished: horticultural successes may be aesthetic failures (and vice versa) and despite momentary perfection – 'that complete aspect of unity and beauty that to the artist's eye forms a picture' – everything changes. In her book *Colour Schemes for the Flower Garden*, Miss Jekyll wrote that 'good gardening means patience and dogged determination. There must be many failures and losses, but by always pushing on there will also be the reward of success. . . .' She was writing at the turn of the century. Half a century later, Victoria Sackville-West (1892–1962) wrote, 'Successful gardening is not necessarily a question of wealth. It is a question of love, taste, and knowledge. . . . Every garden-maker should be an artist along his own lines.'

From the artist-gardener's point of view, white flowers are vital. Both Gertrude Jekyll and Victoria Sackville-West were particularly fond of white flowers and always on the lookout for good white forms. They found them endlessly useful for highlighting shady and dark backgrounds, for complementing pastels, and for relieving colour-saturated plantings.

'Miss Jekyll Alba'

Gertrude Jekyll constantly refers to white flowers and found them indispensable in the gardener's palette. She extolled the virtues of plants as different as the 'white perennial Lupine with an almond-like softness of white', *Lilium candidum* (which 'always holds its own'), *Muscari botryoides* 'Album' ('a capital white'), *Iris unguicularis* 'Alba' ('a beautiful thing') and *Primula japonica* 'Alba' which she found infinitely preferable to the 'rank magenta red' of the type colour. The white form of love-in-the-mist (*Nigella damascena*), an excellent white annual, was deservingly named 'Miss Jekyll Alba'. Her bank of early bulbs consisted largely of drifts of white hyacinths and white crocus between bands of blue scillas, muscari, chionodoxa and pure yellow narcissi. The dappled shade of her woodland garden displayed white flowers at their loveliest, from tiny May lilies (*Maianthemum bifolium*), woodruff (*Galium odoratum*) and chickweed wintergreen (*Trientalis europaea*) to the bolder, more familiar lilies, foxgloves, columbines and willowherb. And where wood and garden met she planted white cistus, clematis and 'Garland' roses which shone bright against the dark interior.

Miss Jekyll's *pièce de résistance* was the herbaceous border – indeed, she pioneered this method of planting which has become the backbone of the English flower garden – and found that the best arrangement of colours was to have pale ones towards the ends and strong ones in the centre. The 'cool ends' consisted mainly of white flowers or a combination of pale yellow, white and palest pink: lilies, *Clematis recta* and *C. flammula* (the latter trained to camouflage the cut-back stems of delphiniums), *Chrysanthemum maximum*, *Gypsophila*, and white forms of foxgloves, campanulas, asters, yuccas, lupins, goat's rue, everlasting peas and dahlias. She grew pots of lilies (mainly *Lilium longiflorum*, *L. candidum* and *L. auratum*) which served as mobile border plants, being dropped in to enhance or enliven neighbouring plants as summer advanced. (*Lilium regale* was not, of course, amongst them as it was only discovered by E. H. Wilson in China a few years before Miss Jekyll's death.) Although complementary colours formed the basis of her colour schemes, she enjoyed the occasional surprise, adding a pot of *Lilium longiflorum* to a hot spot – 'just a flash of their white beauty in the middle region of strong reds'.

The subject of one-colour gardens – 'gardens of restricted colouring', to be exact – always interested Gertrude Jekyll. Like most artists, she found discipline challenging and rewarding. Although she referred to them by their predominant colour (blue gardens, gold gardens, etc.), tone and colour harmony were more important than colour itself. Her critical eye could not bear orange goldfish as an ornamental feature in a gold garden, and she decried those who 'make it a kind of conscience that if it is called a blue garden there shall be nothing in it that is not absolutely blue, whereas the flowers may be praying for the company of white lilies or palest yellow snapdragons or sulphur and white hollyhocks'.

In *Colour Schemes for the Flower Garden*, Gertrude Jekyll describes both a grey garden and a green garden but not a white garden as such. The grey garden consists largely of grey foliage plants with pink, white, lilac and purple flowers, and in the green garden the emphasis is on plants with bright or dark green foliage which is either glossy or fern-like, together with white flowers – campanulas, tulips, peonies, snapdragons, hellebores and, as always, white foxgloves and lilies. From this we may deduce that she found white flowers more effective with foliage of a distinctive green or striking texture, and that perhaps a wholly silver-and-white garden could look a bit faded and dusty without a touch of warmth and dash of contrast.

Of all these so-called one-colour gardens, the white garden is probably the most difficult. The various shades of white have as much to be kept apart as put together, using foliage plants and virtuoso touches to prevent an uninspiring monochrome. With rich greens and interesting shapes, white flowers alone may suffice, but with a filigree of silvery foliage the occasional palest pastel colour may be needed to prevent the eye from roving in search of contrast. The indefinable pinkish-grey of *Campanula burghaltii* or barely pink *Schizostylis coccinea* 'Pallida' are by no means out of place in the white garden, and

plants with arresting shapes – tall flower spikes or big bold leaves – are always welcome.

Sissinghurst's white garden

The most famous white garden of all is the one created by Vita Sackville-West at Sissinghurst Castle in Kent. She wrote that 'It is amusing to make one-colour gardens. They need not necessarily be large, and they need not necessarily be enclosed, though the enclosure of a dark hedge is, of course, ideal. Failing this, any secluded corner will do, or even a strip of border running under a wall, perhaps the wall of the house.' Like Gertrude Jekyll, she did not advocate a slavish adherence to a single colour, but a predominance of one, with the addition of others which are complementary and perhaps a few which make a subtle contrast.

She first thought of a white garden – which she mostly referred to as a grey, green and white garden – in 1939, but it took ten years to materialise. 'It is great fun,' she wrote in the *Observer*, 'and endlessly amusing as an experiment, capable of perennial improvement, as you take away things that don't fit in, or that don't satisfy you, and replace them by something you like better.' Her initial plans, which she visualised as clearly as any artist before a canvas, included a low haze of grey-leaved plants such as southernwood (*Artemisia abrotanum*), artemisias and cotton lavender (*Santolina*), pierced by tall white flowers, especially *Lilium regale* which she grew large numbers of in advance from seed. The aim was 'to have a foundation of large, untroublesome plants with intervening spaces for the occupation of annuals, bulbs or anything else that takes your fancy'.

But Vita Sackville-West was also a poet. Behind the framework of good cultivation and the experiment of 'mixing one sort of plant with another sort of plant, and of seeing how they marry', was a clear idea of its total effect. She appreciated white flowers not just for their beauty and usefulness in the garden, but for their power to induce moods. 'The ice-green shades that it can take on in certain lights, by twilight or moonlight, perhaps by moonlight especially, make a dream of a garden, an unreal vision. . . .' As she planned it, she confided: 'I cannot help hoping that the grey ghostly barn owl will sweep silently across a pale garden, next summer in the twilight – the pale garden that I am now planting, under the first flakes of snow.' Unfortunately, visitors to her white garden now are not allowed the experience of sitting in it as the light fades. They can however, still appreciate its quite different mood on a sunny summer day by pausing under the central arbour of *Rosa longicuspis* when it is in full flower. The huge climber is

Nigella *'Miss Jekyll Alba'* – one of the finest white
annuals, named after Gertrude Jekyll who found
white flowers indispensable.

Platycodon grandiflorus *'Album'* – the white balloon
flower bears inflated buds and huge campanula-like
blooms on fragile growths which disappear
completely in winter.

Cardiocrinum giganteum – *an awe-inspiring Himalayan lily which reaches 10 ft (3 m) tall and smells of honeysuckle and incense.*

Crambe cordifolia *with roses – contrasts of shape and form are important in the white garden.*

supported by metal arches and in its shade sits a dark and massive urn. The effect is surprising – somewhere between entering a cave and emerging from under a bridal veil.

Sissinghurst's white garden was originally a large rectangle divided in half by a path of flagstones over which flopped mats of *Stachys lanata*, *Dianthus* 'Mrs Sinkins' and white pansies. The silver underplanting was not only lovely to look at but also gave shade and shelter to bulbous or slightly tender subjects and covered the bases of bare stems. From it arose the spires of *Eremurus himalaicus*, white delphiniums, campanulas, irises and foxgloves, as well as the royal lily (*Lilium regale*). White lavender, double white primroses and the white balloon flower (*Platycodon grandiflorum* 'Album'), together with pale pink primulas, were added to the edges. Statuesque plants played an important part: clouds of tiny white flowers around the bulk of *Crambe cordifolia*; the giant Himalayan lily (*Cardiocrinum giganteum*) with its tower of massive trumpets; and great silvery Arabian thistles (*Onopordon arabicum*). Over the years, shrubs and climbers took shape: grey-leaved sea buckthorn (*Hippophae rhamnoides*) and *Pyrus salicifolia* sheltered a leaden statue of a vestal virgin; and white-flowered *Hydrangea arborescens* 'Grandiflora', weigela, buddleia, cistus, tree peonies, clematis, and *Rosa longicuspis* added to 'the cool, almost glaucous, effect'. The year ended with a flourish of Japanese anemones and dahlias before the white flowers finally went out like lights among their green and grey setting.

In spite of its creator's fears that it might be 'a terrible failure', Sissinghurst's white garden has gone on from strength to strength, inspiring a host of white beds, borders and other white gardens all over the country. Of course there must have been some failures in the course of its evolution but they undoubtedly received the 'thoughtful care and definite attention' prescribed by Gertrude Jekyll. In Vita Sackville-West's words: 'The gardener must be brutal, and imaginative for the future', '. . . scrap what does not satisfy . . .', 'be quite ruthless about it'. To a certain extent one can safeguard against practical failures by checking the cultural requirements of plants before purchase and, if possible, seeing them as mature specimens or at least when flowering sized. But aesthetic successes and failures are where the real challenges lie, presided over by individual taste and endeavour.

The basics of white gardens

One of the fascinations of white gardens is their subtlety. There are contrasts of course, the obvious being white flowers against dark green foliage, but others are achieved by the emphasis of shape and form rather than the statement of colour: full-blown creamy white roses against a haze of cold white *Crambe cordifolia*; translucent goblets of ivory tulips above flat bright *Anthemis cupaniana* daisies; the filigree silver of *Tanacetum haradjanii* fringing the naked stems of white meadow saffron (*Colchicum autumnale* 'Album').

Planning a white garden involves choosing a framework of key plants which you can fill in as you fancy over the years. The overall structure will probably be dictated by existing features such as walls, hedges, fences and trees. These can be brought into the scheme by clothing them in white-flowered climbers. Red brick walls are prime candidates for white climbers anyway, whether or not they are part of the background for a white garden, all too often suffering the clash of pinks, reds and purples when they could enjoy the flattery of white. If the area is so large and bare or unsightly (or the gardener so impatient) that something is desperately needed to cover it in a season, resist the temptation to plant the Russian vine (*Polygonum baldschuanicum*) unless it can have 40ft (12m) of climbing surface to itself for all time. Instead, try the white form of the half-hardy annual cup-and-saucer vine (*Cobaea scandens* 'Alba') which will easily reach the top of a house wall in the course of the summer and is happy in sun or partial shade. It latches on to most surfaces efficiently by means of branched tendrils though has a tendency to grow ever upward and may need some encouragement to fan out. The pinnate leaves are attractive and the flowers – a large greenish-white bell backed by a saucer-like calyx – are quite spectacular, as are the egg-sized fruits. The seeds are large, germinate rapidly and romp away, so like courgettes they need to be sown under glass in late spring, a few weeks before the end of frosts. The only problem is likely to be that seed of the white form is not easy to come by. Packets of 'mixed' seed are sometimes offered but in my experience consist of mostly the usual purple-flowered sort. 'Alba' seedlings can, however, be told apart from the rest by their bright green leaves which have nothing of the purplish flush that goes with purple blooms.

While the cup-and-saucer vine provides distraction with its effortless advance, more permanent climbers can be patiently established. Among the elite of climbers is wisteria, oriental in origin and distinctly oriental in grace. Its gnarled and twisted branches are counterpoised by elegantly dissected foliage and long vertical trusses of flowers whose perfume adds yet another first to its list of desirable qualities. Wisterias do not have to go against a wall: they are just as successful grown into trees, tortuously trained into arbours or planted in the open as standards. White Chinese wisteria (*Wisteria sinensis* 'Alba') is becoming more and more popular. I recently saw young plants for sale in a supermarket at the price of a rose bush. This species is generally considered finer than its rival, the Japanese wisteria (*W. floribunda* 'Alba'), which tends to have more of a lilac tint.

A succession of white clematis is a great asset, starting with the vigorous *Clematis armandii* which does best in a sunny sheltered position. Unlike most clematis, it has glossy, dark, evergreen leaves which are divided into three long leathery leaflets. The flowers on some plants have a decided pink flush so if you are set on white it is best to choose one in early spring when they are flowering. A good named white, available from specialist growers, is 'Snowdrift', whose flowers are larger than average. *Clematis montana* comes next in the flowering season. 'Grandiflora' is probably the best pure white, and 'Alexander' a good scented creamy white. 'Wilsonii' finishes the montana season, flowering in early summer rather than spring, with an abundance of scented ivory blooms which have unusual twisted sepals. Summer sees avalanches of the large-flowered hybrids. Two of the best are the dark-stamened 'Henryi' and long-flowering cream-stamened 'Marie Boisselot' (syn. 'Madame le Coultre'). 'John Huxtable' has the distinction of being smaller growing and later flowering – a good choice for covering tree stumps. Last in the season is *C. viticella*, a small bushy species which grows well in sun or shade and tends to die back in winter. 'Alba Luxurians' is an interesting form with greenish-white flowers.

Climbing roses are another obvious choice for whitening existing features. 'The Garland' is a nineteenth-century rambler with a twiggy growth and an eventual height of about 15 ft (4.5 m). Its small cream flowers have narrow petals and a rich orangey scent. Rather larger is the old noisette variety 'Mme Alfred Carrière' – hard to beat for vigour, fragrance, length and freedom of flowering and tolerance of shade. It bears bright green foliage, few thorns and globular flowers which change from blush-white buds to cream and then white. 'Paul's Lemon Pillar' may have some shortcomings (it flowers briefly and does not thrive in cold areas) but is one of the most perfectly shaped roses in existence. The blooms are deeply fragrant and the most lovely milky white, shading to a soft greenish-yellow in the shadows, with no hint of pink. Modern climbers include 'Climbing Iceberg', a good weatherproof variety introduced in 1969, about ten years after its debut as a bush rose established it as one of the best of all floribundas.

The most commonly planted climber on shady walls and fences is *Hydrangea petiolaris*. Rather similar but more unusual is *Schizophragma integrifolia*, a self-clinging Chinese species whose summertime display of long white bracts has an arresting grace. It needs to be well fed and mulched for optimum results.

If there is space for a tree near the white garden, the most spectacular of white-flowered species is undoubtedly *Davidia involucrata*, a majestic centrepiece for a large lawn. Descriptively known as the handkerchief tree, dove tree or ghost tree, the sight of its huge pendant white bracts is enough to make you fall to your knees in wonder. The impatient can buy (at a price!) the largest available sapling but it can be grown from seed. The large walnut-like seeds take between eighteen months and two and a half years to germinate, and the tree begins to flower after about seven years.

Magnolias in full flower are almost as breathtaking. A popular centrepiece is *Magnolia* x *soulangeana* which bears waxy tulip-shaped flowers on bare branches in spring. Though usually seen with purple-flushed flowers, there are several pure white varieties. 'Lennei Alba' has very large, rounded flowers. Small enough to take its place in even a modest-sized white border is the slow-growing shrubby *M. stellata*. 'Royal Star' is particularly compact; with narrow-petalled flowers which, as the name suggests, are more starry than goblet-shaped. Rather similar as far as the flowers are concerned is 'Wada's Memory', whose fragrant flowers are produced so extravagantly that they look like petalled clouds against the blue sky. In habit it is quite different and though upright will reach all of 30 ft (9 m). Different again is the evergreen *M. grandiflora* which can be grown as a specimen tree in sheltered gardens or as a wall shrub. Its great ivory goblets have an overpowering lemony fragrance and can perhaps best be appreciated by looking down on them from an upstairs window. If planted beneath a bedroom window, their scent will enter your dreams.

Coming down to earth, there are any number of white-flowering cherries but of particular merit is *Prunus subhirtella* 'Autumnalis', which flowers right through the winter and can be cut for the house whenever you feel the need for a breath of spring. It flowers when little else does, presiding like a guardian angel over the dormant garden while most gardeners are indoors engrossed in seed catalogues and plans for the coming year. Of the spring-flowering varieties, the columnar 'Umeniko' is a sensible choice for restricted spaces, scarcely exceeding 5 ft (1.5 m) across. At the other extreme is 'Shirotae' which spreads its arching branches with the grace of a ballerina, making its debut each spring in a confection of bronze new leaves and scented single to semi-double blossoms.

Less often seen is the supremely beautiful *Eucryphia* x *nymansensis* 'Nymansay', a cross between two Chilean species, the evergreen *E. cordifolia* and *E. glutinosa*, the hardiest *Eucryphia*. Though tolerant of lime (unlike *E. glutinosa*) and

ABOVE Cobaea scandens *'Alba' (cup-and-saucer vine) – a vigorous half-hardy tendril climber which will easily reach the top of a house wall in a season. In its native Mexico, it is pollinated by bats.*

RIGHT Clematis *'John Huxtable' – a late-flowering white cultivar which is excellent for growing over tree stumps and thrives in sun or light shade.*

RIGHT Rosa *'Paul's Lemon Pillar' – one of the most perfectly shaped roses in existence. The blooms are richly fragrant and waxy, with no hint of pink.*

BELOW Schizophragma integrifolia – *a self-clinging Chinese climber with an elegant summertime display of white-bracted flowers.*

quick growing it is of doubtful hardiness in cold areas, especially when young, which no doubt accounts for its relative scarcity in cultivation. It is, however, worth every effort to establish, being smartly evergreen, neatly columnar in shape and covered from top to bottom in large creamy-white stamen-filled flowers in late summer – a time when very few other trees or shrubs are blooming. It also provides a feast for bees which, judging by the numbers, come from miles around to enjoy the floral bonanza.

My shortlist of white-flowered shrubs would have to include at least one mock orange (*Philadelphus*). Where space is not a problem, I would go for 'Belle Etoile' which has large single flowers with a pinkish-maroon centre and a delirious scent. Similar but reaching nearer 8 ft (2.4 m) than 10 ft (3 m) is 'Beauclerk'. One of the smallest is 'Manteau d'Hermine', a double-flowered cultivar which does well even on thin chalk soils and reaches only 3 ft (90 cm). For autumn and winter interest, the evergreen *Viburnum tinus* is easy, reliable and free-flowering, come rain or snow, in sun or partial shade. 'French White' and 'Israel' have whiter than average flowers. It also makes a good hedge or background for white flowers, as does *Osmanthus delavayi*, with much smaller leaves and small tubular scented white flowers in spring. Very similar but hardier and tolerant of dry chalky soil is the intergeneric hybrid x *Osmarea burkwoodii*.

The diva of white-flowered shrubs is *Viburnum plicatum* var. *tomentosum* whose flat clusters of pure white bracts are produced in tier after tier between rows of downward pointing leaves in early summer. The classic variety is 'Mariesii' whose horizontal branches emphasise this cascade effect. Almost as spectacular is the dogwood *Cornus kousa*, another early summer flowerer which is completely covered in pink-flushed white bracts, four to each flower. Both reach 10 ft (3 m) tall and as much across. Another great beauty is the white Spanish broom (*Cytisus albus*, syn. *C. multiflorus*) which is so smothered in milk-white pea flowers that the grey-green stems and minuscule upper leaves are barely visible. It flowers in late spring and is happiest on rather poor soil in full sun, but tends to be short-lived.

One shrub I would perhaps forego in the white garden is *Buddleia davidii*. There are several excellent pure white forms but they have the worst of all habits for white-flowered plants: they retain their dead flowers long after they have turned an unsightly dark brown. The purple varieties do this too, of course, but it goes almost unnoticed. On young bushes it is possible to deadhead them as flowering progresses, but they soon reach a size where this becomes impracticable.

In addition to trees, shrubs and climbers, there are a number of other plants which are bold enough to form the basic framework or landmarks of a white garden. Seakale is a favourite, either *Crambe cordifolia* with dark green leaves or *C. maritima*, the one used as a vegetable, which has glaucous foliage. The former is by far the larger, reaching 6 ft (2 m) with a great cloud of tiny white flowers in summer, but where space is limited or where something grey is wanted, the latter is still quite dominant at only 2 ft (60 cm) and bears generous numbers of creamy-white flowers rather earlier. As a compromise, *C. koktebelica* is midway between the two in size. The great silver thistle, *Onopordon acanthium*, is indispensable too. Though its flowers are purple, the tower of vast down-covered leaves and spiny stems give an overall silver-white effect and are a compelling sight as they reach for the sky during the run-up to flowering. Being biennial it dies after this vast expenditure of energy but is generous with seedlings. Less often seen is the white hellebore, *Veratrum album*, a poisonous plant of the lily family which has very handsome pleated leaves that are a joy in spring, and 4 ft (1.2 m) spires of greenish-white flowers in summer. It prefers moist soil and some shade.

Having planned the main feature plants for in and around the white garden, the next important group to consider is the supporting cast of plants which form mounds and mats. They will provide permanent foregrounds (or backgrounds) for the white-flowered stars – the lilies, aquilegias, lupins, delphiniums and so forth. Some, such as the white-flowered common sage (*Salvia officinalis* 'Albiflora') combine both white flowers and an attractive (and culinarily useful) evergreen velvety mound of foliage. On the whole though, the mound- and mat-formers are dependent on foliage for their effect, the flowers being insignificant. Artemisias are undoubtedly the mainstay of these smaller sculptural plants. 'Powis Castle', a non-flowering hybrid resembling the rather tender *Artemisia arborescens* (which is probably one parent), is particularly good, making a loose finely cut silver mound about 2 ft (60 cm) tall. *A. absinthium* 'Lambrook Silver', the grey version of common wormwood, is slightly smaller. More upright and more green than grey is lad's love, *A. abrotanum*, which like the blue-grey rue (*Ruta graveolens*) benefits from a fairly drastic prune in early spring. The oddly named old warrior (*A. pontica*) comes midway between a mound and a mat, spreading by underground runners into a miniature thicket, 18–24 in (45–60 cm) high, of upright stems clad in finely cut grey foliage. Rather smaller and less tightly packed are the stems of *A. splendens* whose leaves look rather like silver wire.

Mat-forming plants are invaluable at the edges of borders, taking on the role of foothills between the plains of path or lawn and the mountainous clumps of foliage and flowers behind them. Again there are artemisias to do the job. The soft, silvery *A. schmidtiana* 'Nana' makes 4 in (10 cm) hummocks while the ground-hugging *A. stelleriana* 'Nana' has felted chrysanthemum-like leaves and is about twice the height. Furry *Stachys olympica* (syn. *S. lanata*), popularly known as lamb's ears or rabbit's ears, is a perennial favourite though its mats quickly become carpets. The variety 'Silver Carpet' (a non-flowering clone) is aptly named. Equally enthusiastic in sunny positions is *Anthemis cupaniana*, with grey ferny foliage and an extrovert display of large single daisies in early summer. It is an excellent plant for impatient gardeners, looking as if it has been there for years after only a few weeks, but needing a regular trim if it is not to invade more sensitive newcomers who are still tentatively putting down roots.

For semi-shaded areas of a white garden, an edging of lungwort gives a long period of interest. There is a white form of common lungwort, *Pulmonaria officinalis* 'Alba', but better still is 'Sissinghurst White', which has rich green white-spotted leaves that remain eyecatching after the springtime show of white bells has faded. Dwarf comfrey (*Symphytum grandiflorum*) is another good, if rather rampant edger in shade, with an offering of cream bells in early spring as the new leaves emerge. The variegated form is more interesting and less vigorous. Most silver edging plants need full sun but *Lamium maculatum*, a creeping dead nettle, is quite content in partial shade. There are several white-flowered forms, the best being 'White Nancy'.

Once the permanent features of the white garden are established – trees, shrubs, climbers, statuesque plants and various islands and margins of foliage interest – it becomes much easier to position clumps and spires of perennials and to find homes for bulbs which can emerge through the canopy of vegetation. Then, when growth gets under way, remaining gaps can be filled in with annuals and half-hardy subjects.

The beauty and value of white flowers does not, of course, need a white garden for their appreciation. There are situations in any garden which may cry out for something white – dappled woodland, heavy shade, water, sun-drenched patios, arbours and conservatories. Areas with special problems or potential, or those you spend more time in, are worth indulging.

White cottage-garden flowers

Some of the loveliest white flowers are the old-fashioned ones. They are as clean and pretty in effect as linen and lace, an uplifting sight on their own or delightful beside other colours. A great favourite of mine is white musk mallow (*Malva moschata* 'Alba'), a well-behaved border plant which rarely gets too tall that it needs staking, has finely cut foliage and will flower throughout the summer. It spreads and sows itself with similar decorum and its dead flowers politely disappear when their hour of glory is over. For a damper spot, the double form of fair maids of France (*Ranunculus aconitifolius* 'Flore Pleno') is a delight for three weeks in late spring, its branched stems bearing numerous small white button flowers. It was probably introduced by Huguenot refugees who brought their favourite flowers with them as they fled from France in the seventeenth century. Again, it is the model of deportment and should need no staking.

In contrast, white goat's rue (*Galega officinalis* 'Alba') is anything but tidy, though an easy and lovely border plant for all that with spikes of tiny white pea flowers for weeks in the summer. *G. hartlandii* 'Candida' is very similar but more densely flowered and upright. Even more floppy is soapwort *Saponaria officinalis* 'Alba Plena' whose double white flowers look like miniature carnations and smell lovely too. They flower very late, giving them their country name of goodbye-to-summer. Soapwort is a long-lived plant. It was once grown not just for its beauty but also for its chemistry. The sap contains soap-like substances which were used for washing clothes before bars of soap were a household item.

The tail-end of summer also sees the tidily upright obedient plant (*Physostegia virginiana* 'Alba'), the curiously crooked spikes of *Lysimachia clethroides* and faultless Japanese anemones (*Anemone* x *hybrida*) blooming as if it were spring. Though only introduced in the nineteenth century, Japanese anemones soon became firmly established in the repertoire of cottage gardens. The first white form, 'Honorine Jobert', remains one of the best.

For sheer dignity, peonies are unsurpassable. The old double white *P. officinalis* 'Alba Plena' is difficult to find now but the nineteenth-century double white 'Duchesse de Nemours', a fragrant *P. albiflora* variety, is readily available. Single forms of *P. albiflora* are, if anything, even lovelier. One of the earliest to grace our gardens was 'Whitleyi Major', which Reginald Whitley, a Fulham nurseryman, introduced from Canton in 1808. Some catalogues still list it.

No border is complete without phloxes, which raise their bright heads and give off a peppery scent from midsummer until early autumn. The old-fashioned pure white *Phlox paniculata* 'Alba' is one of the most stately. The species was introduced from the United States in the early eighteenth century and can reach 4 ft (1.2 m) so is a candidate for the back of large borders. *P. maculata* 'Miss Lingard', another old variety, is earlier and smaller.

ABOVE Prunus subhirtella *'Autumnalis' – a winter-flowering cherry which blooms for months and can be cut for the house whenever you feel like a breath of spring.*

RIGHT Philadelphus *'Belle Etoile' – a highly fragrant mock orange with large purplish-centred single flowers.*

RIGHT Viburnum plicatum *var.* tomentosum *'Mariesii' – the diva of white-flowered shrubs.*

BELOW Cytisus albus *(syn.* C. multiflorus*) – the white Spanish broom is a spectacular if rather short-lived shrub for poor soil in sun.*

Hardy geraniums have long been popular as garden flowers. The herbalist John Gerard grew *Geranium phaeum* in the sixteenth century, since when it has become established in the wild. Its white form is quite lovely, as is that of the meadow cranesbill (*G. pratense* 'Albiflorum') which also has a double white form 'Plenum Album'. Considered by many as the best of all white hardy geraniums is the one commonly known as 'Rectum Album' (now *G. clarkei* 'Kashmir White') whose perfect saucer-shaped white flowers are veined with lilac. Even the white form of the humble herb Robert (*Geranium robertianum* 'Album') is well worth growing. A dwarf green-stemmed form (usually the stalks are brown) called 'Celtic White' is a collector's item. From a foliage point of view, the best is *G. renardii* with neatly cut and embossed velvety leaves of a soft sage green. From a distance its flowers look pale pink but a closer look reveals the kind of detailed beauty we have come to expect from this group, the pink colouring really a network of delicate veins on notched white petals. Of all hardy geraniums, *G. macrorrhizum* must be the most useful, with creeping clusters of pungent-smelling leaves which make excellent ground-cover or edging in sun or partial shade. The white form ('Album') is just as pretty as the pink.

Many cottage-garden flowers were originally grown for medicinal purposes as well as for their attractive flowers. Just as useful and even prettier are the double forms, such as double feverfew (*Tanacetum parthenium* 'Plenum'), double chamomile (*Chamaemelum nobile* 'Flore Pleno') and double sneezewort (*Achillea ptarmica* 'Flore Pleno') (now superseded by 'The Pearl'), all with white pompom daisies – not to mention the double daisy itself (*Bellis perennis* 'Alba Plena'). Dwarf double-flowered varieties of feverfew ('Snow Dwarf', 'White Gem', 'Tom Thumb White Stars' and so forth) are well worth growing as half-hardy annuals for edging, bedding, containers or filling gaps in borders, producing a compact mound of neat chrysanthemum-like foliage and masses of long-lasting flowers, even in dry shade. The smallest is 'Golden Moss' which produces mounds of yellow foliage and single daisies only 4 in (10 cm) high. All are easy from seed (commonly listed under *Matricaria*) and often sow themselves.

Unusual forms of wild flowers, whether useful or not, have always been a source of cottage garden flowers. The campions make excellent garden plants, among them the white forms of ragged robin, *Lychnis flos-cuculi* 'Albiflora', and the sticky catchfly *L. viscaria* 'Alba' (from northern Europe), the Maltese cross *L. chalcedonica* 'Alba' (a Russian species), *L. flos-jovis* 'Alba' and grey-leaved *L. coronaria* (from southern Europe) – the latter being a particularly good plant for silver gardens. Some of the willowherbs have proved worthy of cultivation too. In Victorian times the great or hairy willowherb (*Epilobium hirsutum*) – known in country areas as codlins-and-cream after the stewed apple smell of its leaves – was sometimes seen in gardens. It is tall – up to 5 ft (1.5 m) – and fairly invasive but its white form would be attractive in damp areas of the wild garden. William Robinson mentions a variegated form in *The English Flower Garden* (1883) and also the white form of rosebay willowherb, *E. angustifolium* 'Album', a plant which is rapidly gaining popularity once more. A fine stand of this in the Chelsea Physic Garden recently drew more attention and admiration than everything around it. Much smaller is *E. glabellum*, a relative from New Zealand which has creamy-white flowers and is good at the front of borders.

Granny's bonnets or columbines (*Aquilegia vulgaris*) are, as their names suggest, archetypal cottage-garden flowers. Both white and double white forms were grown in Victorian times, leaning out from the shade of hedges and shrubs where they had sown themselves. Traditional, too, were hollyhocks (*Althaea rosea*) which stood like sentinels beside doors and windows with their backs to the warm stone wall. There was once a great range of named varieties, from almost black to glorious pure whites, like 'Alba Superba', 'Cygnet' and 'Queen of the Whites', and more that were pastel-tinted, such as 'Princess', with a hint of salmon, and the lavender-white 'Mrs Elliott'. Unfortunately, the dreaded rust put paid to them and in their place we have hybrids of mixed colours, including white. Though hardy perennials, devotees can now grow them as half-hardy annuals and weed out unwanted colours as they begin to open.

Flag irises (hybrids of *Iris germanica*) and a great variety of pinks or clove gilliflowers (*Dianthus* species) have always been popular. Again, there are a number of fine white cultivars. Old-fashioned pinks are undergoing a revival of interest and many cottage-garden varieties are in circulation once more. 'Mrs Sinkins' remains a firm favourite, with deliciously scented flowers whose fringed petals dissolutely burst out of their bodices, a habit now regarded as endearing though once considered a great failing. Most of the flag irises grown today are recent hybrids. By all accounts, the old varieties were not a patch on them. There are many good whites – tall, intermediate and dwarf. 'White City' is an older one, dating back to 1940, with smallish blue-white flowers on medium-sized plants. 'Lilliwhite', made in 1958, is probably the best white dwarf bearded cultivar, with ruffled falls and beautifully shaped pure white flowers. Though rather jazzy, 'Frost and Flame' is an outstanding variety with bright orange beards which emphasise the whiteness of the petals. Less obtrusive is the tall, immaculate 'Cliffs of Dover'.

The classic cottage-garden shrubs are surely roses and lilac (*Syringa vulgaris*). The latter was recorded in Henry VIII's garden, having been introduced from Persia in the sixteenth century. In Victorian times a white variety known as 'Alba Virginalis' was recommended for forcing, presumably being pot-grown in conservatories for the purpose. There are a number of single and double whites available now, both pure white and ivory, with little to choose between them in quality of blooms or fragrance. Lilac needs an open sunny position. If crowded by trees and shrubs the wood fails to ripen and it becomes all leaf and no flower.

The roses display much more variety in habit, foliage and flowers. *Rosa* x *alba* 'Semi-Plena', the white rose of York, is a must with its grey-green leaves and richly fragrant semi-double blooms. 'Maxima', the Jacobite rose, has double flowers, blush-white in bud, and a redoubtable constitution. Albas date back to the Middle Ages. They are extremely hardy, virtually disease-free, scarcely need pruning and thrive in partial shade, coastal areas, stony soils and similarly challenging conditions. Despite their name, they are not necessarily white. In fact, apart from the two just mentioned, most are pink. Damask roses are mostly pink too; the best white probably being 'Mme Hardy', slightly lemon-scented and with flat, fully quartered blooms which are all but perfect in shape. Cabbage roses (also known as Provence or centifolia roses) and moss roses are quintessential for scent and sumptuousness. Centifolias are lax and thorny with the globular blooms that feature in paintings by the old Dutch Masters. 'Blanchfleur' is a vigorous white nineteenth-century variety with red-tipped buds. 'White de Meaux' is smaller and more compact, seldom exceeding 3 ft (90 cm). Moss roses originated as mutations of *R. centifolia* and were great favourites in Victorian times. They are characterised by resin-scented moss-like growths around the stalks and sepals. One of the best known is 'White Bath' (also known as 'White Moss') with pure white, deeply fragrant flowers on an average-sized bush. 'Blanche Moreau' is rather taller, reaching 6 ft (1.8 m) with brown moss and rather globular creamy-white blooms. Equally tall but more vigorous is the lovely 'Comtesse de Murinais', whose flowers open flat and blush-white but are pure white and green-centred when full-blown.

Part of the charm of the cottage garden is its crowded, multi-layered effect. Angles are rounded and contours blurred in ground-smothering vegetation and a tumble of flowers. Trees, shrubs and perennials are skeined together with clematis and everlasting peas (*Lathyrus latifolius*) – the latter having an excellent white form, 'White Pearl' – and a profusion of pinks, violas, lavender and catmint spills over the path. Staking never quite succeeds and wands of campanulas, larkspur and love-in-the-mist stray among the more determinedly upright dahlias, yuccas and cannas. For a white garden to work well, it has to emulate this kind of profusion. A few white dots here and there are neither here nor there. Treasures there may be, but in the main floriferousness combined with a reasonable length of flowering is what is needed. Fortunately there are white forms of many cottage-garden flowers that will achieve that kind of abundance. Many have been mentioned already in various contexts. To them should be added a final word of recommendation for: *Viola cornuta* 'Alba', which will flower in sun or shade from early summer to autumn; the white form of red valerian (*Centranthus ruber* 'Albus'), which will grow in the worst soil imaginable, as well as out of walls; champion annuals *Nigella damascena* 'Miss Jekyll Alba', *Lavatera trimestris* 'Mont Blanc', *Cosmos bipinnatus* 'Purity' and *Cleome spinosa* 'Helen Campbell'; *Leucanthemum maximum* (syn. *Chrysanthemum maximum*) 'Beauté Nivelloise' and 'Phyllis Smith' – large, strong, long-serving daisies; reliable campanulas of every description from giant *C. latifolia* 'Alba' to elegant *C. alliariifolia*; and white dahlias of any description, from formal balls to neat decoratives and exuberant cactus-flowered varieties, which witness the dying of summer with such superlative confidence. Stalwarts like these hold the stage set by the framework of dependable climbers, trees, shrubs and carpeters and, with a supporting cast of bulbous plants, will provide long periods of interest while the stars – tree peonies, delphiniums, eremurus *et al.* – come and go.

Round the year with white-flowered bulbs

It is, of course, impossible to have a white garden filled with flowers all year but generous plantings of bulbs around dormant perennials and under ground-cover will go some way to achieving at least year-round interest. They are among the first flowers of the year, and the last, beginning with snowdrops and crocus and ending with colchicums (often mistakenly called autumn crocus), cyclamen and autumn-flowering crocus.

No garden, white or otherwise, is complete without snowdrops. The most commonly planted species is *Galanthus nivalis* and its double form 'Plenus', but there are a surprising number of other species and varieties for the enthusiast. The hybrid 'S. Arnott' is an absolute must: vigorous, with huge beautifully shaped and scented flowers. Fragrance is not something we usually associate with snowdrops (and it may require that you go down on bended knees or pick them to

LEFT Salvia officinalis *'Albiflora'* – *the white-flowered form of common sage is also a useful culinary herb and attractive aromatic evergreen.*

BELOW Ranunculus aconitifolius *'Flore Pleno' (fair maids of France)* – *the perfect companion for white-variegated hostas.*

ABOVE Lysimachia clethroides – *a Japanese loosestrife which produces curiously crooked spikes of tiny white flowers in late summer.*

RIGHT Anemone *x* hybrida *'Honorine Jobert' – from late summer to autumn faultless Japanese anemones bloom as if it were spring.*

appreciate it, as scents do not travel far in cold air), but if even the thought of it appeals, try to find 'Brenda Troyle' too. This hybrid has very large rounded flowers which have a pronounced perfume. Incidentally, patience is a virtue as far as snowdrops are concerned. They must have the right conditions – moist soil and summer shade – and dislike disturbance. Buy them 'in the green' if possible (i.e. in the spring when they still have leaves) but even then do not expect great things for a year or two after planting.

Hard on the heels of snowdrops come a whole host of early spring bulbs. Those I would not miss for the world include: the long-lasting floriferous *Crocus chrysanthus* 'Snow Bunting'; along with white hyacinths ('L'Innocence' being an all-time favourite), grape hyacinths (*Muscari botryoides* 'Album'), *Anemone blanda* 'White Splendour' and *Iris reticulata* 'Natasha' (the nearest yet to a white reticulata). As spring advances, bulbous plants seem to be making a take-over bid for the garden. Even the dullest suburban patch is suddenly awash with primary yellow forsythia and matching daffodils. The growing popularity of white daffodils may do something to halt this tidal wave of carnival colours which make such a forceful rebellion against the muted shades of winter. How about a more modest throng of the elegant white cyclamineus hybrid 'Jenny' or classics such as the white trumpet 'Mount Hood'? They look particularly fine near the red winter stems of *Cornus alba* and show up just as well from a distance. For containers or somewhere that can be viewed at close quarters, there is nothing prettier than *Narcissus triandrus* 'Albus', a diminutive species which has been christened angel's tears. The last and perhaps loveliest of all narcissi is *N. poeticus*, the poet's narcissus or pheasant's eye. Its pure white flowers with their small red-edged yellow centres, are simply beautiful and sweetly scented. 'Actaea' is the largest of its kind. There is a double form ('Flore Pleno') which seems to miss the point.

Tulips are almost as popular as daffodils and narcissi but for the most part are nowhere near as permanent. The majority of *Narcissus* species and hybrids will increase year after year, even in grassland and semi-wild areas of the garden, but hybrid tulips rarely do anything the second year and have to be regarded as annuals. Indeed, leaving them in the ground may encourage diseases that can ruin future plantings. One of the few white-flowered tulips which will settle down and flower annually is *T. biflora*, a small species that produces several starry yellow-centred white flowers per stem. But in spite of their evanescence, hybrid tulips are indispensable: theatrical, imperious creations with a formal splendour unmatched by any other spring-flowering bulb. Any white tulip is worth having, but especially recommended are the lily-flowered 'White Triumphator', oval 'Snowpeak', ornate 'White Parrot' together with the white form of *T. fosteriana* ('White Emperor') and the old variety 'White Victory', which smells like jasmine.

The white-flowered bulbs so far mentioned are very familiar and for the most part are very easy to grow. There are some others, however, which are well worth trying but which need rather special conditions. Enjoying the same moist soil and summer shade as snowdrops and looking rather like an absurdly tall one, the summer snowflake (*Leucojum aestivum*) flowers in spring, in spite of its name. Apart from these basic requirements, it is easy and long-lived, with clusters of rounded snowdrop-like flowers on 18 in (45 cm) stalks. 'Gravetye Giant' is the best variety. The snakeshead fritillary (*Fritillaria meleagris*) likes damp, even heavy soil, in sun. It has two white forms: 'Alba', with green-veined white flowers; and 'Aphrodite', whose flowers are whiter. Trilliums are usually counted as bulbous though they are mostly rhizomatous. They need woodland conditions in moisture-retentive humus-rich soil. The best and easiest white-flowered species is *T. grandiflorum*, which also has a double form, 'Flore Pleno'. At the other extreme is the Algerian *Iris unguicularis*, a winter-flowering species which needs a hot dry sunny spot. In cold areas a wall will act as a night storage heater to guarantee sufficient warmth. After a good summer it can flower from autumn through to spring. The white form, 'Alba', is rather rare.

We tend to think of bulbs in terms of spring flowers but many give their display in summer and surprise us with glorious blooms as the weather deteriorates in the autumn. Alliums are good value for light well-drained soil. Some are normally white-flowered; others have excellent white forms; they may make good cut flowers and dry well; a few are onion or garlic-flavoured and edible. Several of the most popular species have pink to purple flowers and although they do not appear to have white forms recorded, may well produce them sooner or later. Only recently I spotted a lone white-flowered chive (*A. schoenoprasum*) amongst the purple at a herb nursery. Hopefully it may now be propagated and made available. Meanwhile, I eagerly await the discovery of an albino *A. giganteum*!

The best-known white-flowered *Allium* species is *A. neapolitanum*, which produces 2 in (5 cm) umbels of star-shaped flowers on 12 in (30 cm) stalks in spring. Rather similar but later, better and slightly taller is *A. cowanii*. *A. pulchellum* 'Album' is probably the best of all, with stiff 15 in (38 cm) lollipops of white bells at the end of summer. For sheltered conditions in dappled shade – perhaps under wall shrubs – *A. triquetrum* is a good choice, spreading into ground-cover when happy. It has distinctive three-sided stalks and flowers from late spring to early summer. Tougher, tidier but much more invasive is ramsons (*A. ursinum*). It is a lovely sight (if not a lovely smell) when in flower, with its broad dark-green leaves, which reek of garlic, and heads of white stars. In view of its manners, it should be confined to damp shady areas of the wild garden or the banks of a stream.

Lilies are undoubtedly the most spectacular of bulbous plants. Some make better garden plants than others. *L. longiflorum*, *L. auratum* and *L. speciosum* 'Album' cannot be risked in the open ground in cold areas. Of those that are reasonably reliable, *L. martagon* 'Album' has everything to recommend it but its scent, which is faintly unpleasant. It is exceptionally vigorous for a white form, tolerates lime and spreads well in the border or in woodland conditions when established. Its ivory turk's cap flowers are much prettier than the dusky purple of the species and show up better in shade. *L. regale* is a superlative border plant. The species has a maroon flush on the outer petals but 'Album' is very nearly pure white. It increases quickly in most soils and its liking for shade around the roots but a sunny position makes it at home among low mounds of artemisias.

The beloved Madonna lily, *L. candidum*, is unfortunately much more temperamental. In the wild it grows in well-drained, often calcareous soils in rocky places with the bulb near the surface. Its lifecycle is rather unusual. Unlike most other lilies which are completely dormant in winter, it forms a rosette of leaves in the autumn which overwinter, elongating to a metre or more during the spring, flowering in early to mid-summer and becoming briefly dormant in August/September. The blissfully scented, pure white trumpet-shaped flowers are borne in terminal heads of up to twenty, the upper ones more or less erect and the lower ones held horizontally or pointing slightly downwards. Though once extremely common in gardens of northern Europe, it now appears to be suffering a serious decline. In addition to being unpredictable in cultivation – thriving on heavy clay and neglect in one garden, and obstinately refusing to grow in the same or better conditions in the next – it succumbs to virus and botrytis and rarely sets seed. (It is a virgin lily in more ways than one, and has produced only one hybrid – the nankeen lily, *L.* x *testaceum*, when in a momentary lapse of sterility it crossed with *L. chalcedonicum*.) Another problem is that it is difficult to establish and resents disturbance – or so we are told. However, I once planted a bulb which sure enough did nothing but sulk for years, in spite of its idyllic cottage-garden setting. However, after being dug up one August during house removal, it suddenly grew large and lush when it should have

sulked more resolutely than ever, producing two stems of flowers the next summer in suburban soil I would not wish upon a weed. It is indeed a capricious plant.

Late summer sees the flowering of several species of large bulbous plants. The southern African summer hyacinth, *Galtonia candicans*, reaches up to 4ft (1.2m) tall and produces loose spikes of twenty or more white bells from mid-summer to early autumn. Though it has a tendency to peter out in some gardens, perhaps weakened by virus or grey mould, it also seeds itself quite freely. It is best planted in bold clumps and seems to prefer slightly acid soil. *Agapanthus orientalis*, *Amaryllis belladonna* and *Crinum* x *powellii* also need plenty of room. All have white forms but none are reliably hardy. Best is *Agapanthus orientalis* 'Albus', an ideal summer-flowering subject for large pots which can be given protection in winter and whose foliage is a good deal tidier than that of the other two mentioned.

The tail-end of summer and first inkling of autumn are made bearable by such delights as tiny *Cyclamen* and autumn-flowering *Crocus* species. The easiest and most floriferous white cyclamen is *C. hederifolium* (syn. *C. neapolitanum*) 'Album'. Its pure white flowers are often larger than the usual pink form and are produced from late summer to early winter. The foliage is very variable but whether dark green with silver marbling or almost entirely silver, it makes good ground-cover under trees and shrubs from autumn to spring. *C. coum* 'Album' is rather similar but later, flowering from the dead of winter to early spring. The leaves are usually plain green and the flowers dumpier in appearance than those of *C. hederifolium*. Pure whites have been recorded but are extremely rare: the usual white form has a pink nose.

The best white autumn-flowering crocus is *C. ochroleucus*, a Syrian species which produces its delicate flowers in even the worst autumnal deluges. Tempting, though totally unreliable without a summer baking, is the white saffron crocus, *C. sativus* var. *cartwrightianus* 'Albus'. The disfigurement of rain splashing can be avoided by planting the bulbs beneath creepers of thymes and acaenas. Much larger are the so-called autumn crocus, more correctly known as meadow saffron (*Colchicum* species). *C. speciosum* 'Album' is particularly fine, with substantial goblet-shaped flowers. Smaller and more fragile but very free-flowering is the white form of *C. autumnale*. Its double form, 'Album Plenum' is an unusual, if expensive, addition to the collection. Colchicums needs moist soil which does not dry out in summer and shelter from wind for their delicate blooms. They should certainly not be planted beneath carpeters as their large cabbagey leaves, which come up in spring, would smother any fine ground-hugging foliage.

ABOVE Geranium phaeum *'Album'* – *the white form of the mourning widow, a hardy geranium which thrives in shade.*

LEFT Epilobium hirsutum *'Album'* – *the white form of the great willowherb (known also as codlins-and-cream because of the stewed apple smell of its leaves) makes an interesting plant for damp places in the wild garden.*

ABOVE *Viola cornuta 'Alba' – the white horned violet thrives in sun or shade and flowers from early summer to late autumn.*

RIGHT Cosmos bipinnatus *'Purity' – one of the best white-flowered annuals.*

Spotlight on white

One might expect that white flowers, with their virtual absence of colour, would be neutral in effect. Admittedly white flowers will pale into insignificance if unwisely positioned against a white wall, but against dark green or a shaded background, white flowers are certainly not neutral, but provide a most dramatic contrast. Like any colour, white is receptive to changes in light. In fact, it is perhaps *the* most receptive. White flowers are particularly striking when backlit by low morning or evening sun (though from a photographic point of view they will then appear pinkish-gold) and when sunlit against a dark background. The first time I appreciated this was when by chance I planted a white peach-leaved campanula (*Campanula persicifolia* 'Alba') at the edge of a path which led to a well – a fine feature with an old ridged and tiled roof, backed on the south side by a mighty laurel hedge which kept the well itself shaded all day. In some ways it was a rather sinister spot where grass snakes lay in wait for unsuspecting frogs that were drawn to the cool damp void. But in the morning on sunny days in June the view was transformed, the black interior serving as a screen against which were projected the translucent brilliance of the campanula's bells. It is therefore worth looking round the garden on a sunny morning or evening for focal points which remain in heavy shade when the rest is bathed in light. Planting something with white flowers in these positions will highlight this effect, campanulas and foxgloves being especially good as the light shines through their slender growths and illuminates bell-shaped flowers most effectively.

White in shade

Following on from this, white is the best lightener of shade, even in areas which receive no direct sun at all. One of the finest shrubs for shady corners is the white-variegated elder, *Sambucus nigra* 'Marginata' (also known as 'Albovariegata' or 'Albomarginata') which has cream-margined leaves. Almost as good is the golden-variegated 'Aureo-marginata'. Less common is 'Pulverulenta', another white-variegated form but this time with white mottling, which is quieter in effect than the bright margins of its cousins. All are laden with cream flowers in early summer. Elders thrive in damp gloomy spots but do equally well in sunny positions. They are deciduous and should be cut back hard in late winter to encourage strong new colourful growths. If neglected, they try to become trees and the attractive foliage is lost to sight above a jumble of woody stems.

Where something smaller and neater is required, the dwarf evergreen cherry laurel (*Prunus laurocerasus* 'Otto Luyken') is hard to beat. It is tough and tolerant, semi-erect in habit with narrow shiny leaves and upright spikes of white flowers in spring. Like elder, it copes with a wide range of soils and conditions, including sunless areas.

Areas around trees and shrubs are some of the shadiest places in the garden. White-flowered ground-cover is particularly effective where it can be established. To give it the best chance, plant in autumn or early spring when deciduous trees and shrubs are leafless and the ground beneath them is receiving more light than in the rest of the year. White-flowered bugle (*Ajuga reptans* 'Alba') has evergreen rosettes of shiny dark-green leaves which quickly spread into a mat, punctuated by spikes of white flowers in spring. It tolerates most soils and conditions from fairly dry to decidedly wet. Evergreen also is lesser periwinkle (*Vinca minor*). This well-known ground-cover plant is persistent and reasonably tolerant, though it is not happy in wet places and generally tends to be rather sparse. It is a trailer, with strings of small elliptic leaves which root at intervals and bear a scattering of propeller-shaped flowers in spring. There are several white-flowered forms: 'Alba', the golden-variegated 'Alba Variegata', and the unusual 'Gertrude Jekyll' with very small leaves and a clump-forming habit. A double white form, 'Plena Alba', was grown in the eighteenth and nineteenth centuries but is now exceedingly rare or even extinct.

Another good evergreen ground-coverer in deep shade is the sub-shrub *Pachysandra terminalis*, a relative of box (*Buxus*). It forms a thicket of short stems clothed in almost diamond-shaped leaves, topped with white-stamened flowers in spring. It copes well with dry shade. The variegated form 'Variegata' is brighter but less vigorous. On acid soils creeping dogwood (*Cornus canadensis*) is the best choice, forming shrubby carpets which are attractively studded with white flowers (actually tiny greenish flowers surrounded by four showy white bracts) in spring, followed by red berries in autumn.

Woodruff (*Galium odoratum*) is one of the most delicate (in effect) and robust (in constitution) of woodland ground-coverers, though it disappears completely in winter. Preferring moist alkaline soils, it makes large patches of whorled foliage topped by clusters of tiny white stars in spring. Variegated ground elder (*Aegopodium podagraria* 'Variegata') is even brighter in effect with its showy white-variegated leaves. It is not, however, so happy in dense shade, nor are the flowers (which are produced in umbels during the summer) particularly attractive. Nevertheless, in the right spot it is superlative deciduous ground-cover, vigorous but without the dreaded rampageousness of ordinary green ground elder. It could be underplanted with early spring-flowering bulbs.

Few white-flowered annuals or perennials will thrive in unremitting shade. Most appreciate a dappling of sunlight or at least some sunshine in winter and early spring during their main spurt of growth when deciduous trees and shrubs are bare. One of the few annuals which can be recommended is spring beauty (*Claytonia perfoliata*, syn. *Montia perfoliata*), an intriguing if not showy plant with pairs of leaves fused round the stalk like little plates on which sit posies of minute white flowers. Where happy – in neutral to slightly acid soil – it usually sows itself merrily. There are masses of it under trees in the Royal Botanic Gardens, Kew.

Hostas and bergenias are best among shade-loving bold perennials whose foliage is as much a feature as their flowers. *Hosta* 'Royal Standard' and *Bergenia* 'Bressingham White' are exceptionally good, both with magnificent leaves and generous numbers of pure white flowers (scented in the case of the hosta). Rather slow to establish but unique in its lobed evergreen leaves and winter flowering is the Christmas rose, *Helleborus niger*. An ideal position would be at the foot of a wall where its lovely white blooms can be protected from splashing by a lean-to pane of glass. It is best to start with young plants removed carefully from their pots as older ones are most resentful of root disturbance.

Rather more delicate in effect are *Geranium phaeum* 'Album' (the white-flowered form of the mourning widow), an exceptionally adaptable and charming hardy geranium which takes dry shade in its stride; and the little-known *Epimedium* x *youngianum* 'Niveum', a colourful plant with pretty white flowers in spring above lopsided heart-shaped leaflets which have prickly teeth. The foliage is attractively veined and flushed red when young and assumes reddish tints again in autumn. For sheer elegance, the dicentras are outstanding with their fern-like foliage and locket flowers. *Dicentra eximia* 'Alba' has grey-green leaves and clusters of white flowers in late spring and at intervals until the end of the summer. *D. formosa*, very similar but with fresh green foliage, also has a good white form. Even lovelier is the white bleeding heart, *D. spectabilis* 'Alba'. It is larger all round, with the most beautiful divided leaves and arching sprays of huge glistening lockets. Unfortunately, it is rather sensitive and sometimes either fails to appear in the spring or gets its new foliage badly frosted. This is, however, the kind of plant to indulge in, heaping protection and care upon it in hard weather and, just in case, buying another each spring. If you have a cool conservatory, it is worth keeping in pots for a while. There they will produce their fragile leaves and jewel-like flowers for several months, more like something from a tropical paradise than a shady border.

Two of the most striking tall plants for shade are Solomon's seal (*Polygonatum multiflorum*) and false Solomon's seal (*Smilacina racemosa*). As their common names suggest, they look rather alike, at least when not flowering, with pairs of ovate to lanceolate leaves along stiff stems. Their flowers are quite different though: those of Solomon's seal being like waxy white beads dangling under the arching stems, in contrast to those of the false Solomon's seal, which are tiny and borne in terminal plumes resembling lily-of-the-valley in fragrance. They also differ in the kind of soil they enjoy: the former preferring slightly alkaline conditions; the latter being happier where it is on the acid side. There is a variegated form of Solomon's seal which is quite handsome. The closely related species *P. odoratum* (angular Solomon's seal) is less often seen. It is smaller and slower growing and the flowers are scented. It has both white- and gold-variegated forms ('Variegatum' and 'Gilt Edge' respectively). Similar in foliage, though unrelated and quite different in flower is the willow gentian (*Gentiana asclepiadea*). Its willowy stems bear perfect gentians – pure white in the 'Alba' form – in late summer.

Where something really small is wanted for a shady corner, what could be nicer than white violets (*Viola odorata* 'Alba') and primroses (*Primula vulgaris* 'Alba')? Only, perhaps, double white primroses ('Alba Plena') or the choice double white jack-in-the-green 'Dawn Ansell' whose flowers are backed by a ruff of miniature leaves. These treasures flower when the garden can be a cruel place with hard frosts, snow or cold and heavy rain after the mildest spell, so it is tempting to rescue one or two for pots in the conservatory so that they can flower in perfection and be enjoyed to the full. I even dig up chunks of the white form of blue-eyed Mary, *Omphalodes verna* 'Alba', which flowers so delicately at the first peep of spring. In good peaty soil it can be classed as ground-cover and is a delight when sprinkled with its pure white forget-me-nots. Other fragile wonders for the front of shady borders are the white varieties of *Primula sieboldii*. Both leaves and flowers are more delicately cut than those of the common primrose.

The best biennial for shade has to be the white foxglove, *Digitalis purpurea* 'Alba'. Happiest on slightly acid soils, damp or dry, it will sow itself serendipitously year after year (though producing the odd purple-flowered plant) and looks splendid among shrubs (including old roses), under trees, in the wild garden or at the back of herbaceous borders. The seed needs light to germinate so should just be sprinkled on the surface. Foxgloves are at their peak in summer, while the next best – white honesty (*Lunaria annua* 'Alba') – is a spring and autumn plant. In spring its coarse heart-shaped leaves are thankfully hidden beneath branched racemes of chalk-white four-petalled

ABOVE LEFT Cleome spinosa *'Helen Campbell' – a fine white form of the tropical American spider flower, an outstanding half-hardy annual for sunny borders or large pots in the conservatory.*

ABOVE RIGHT Digitalis purpurea *'Alba' – slender plants with bell-shaped flowers are particularly effective in semi-shade. When caught in a shaft of light, they become translucent against the shadowed background.*

BELOW *White tulips – evanescent but indispensable bulbs for the white garden.*

RIGHT Lilium martagon *'Album' – superior in every way to the usual pinkish-purple form, this vigorous reliable Turk's cap lily has everything to recommend it but its faintly unpleasant scent.*

BELOW Galtonia candicans *– bold clumps of the summer hyacinth make a fine show from mid-summer to early autumn.*

flowers. After flowering, the whole plant is forgotten as surrounding perennials jostle for attention, only to re-emerge in the autumn with branches of silvery discs as the competition thins out. Even better than 'Alba' is the form with both white flowers and white-variegated leaves, 'Variegata'. It makes a good companion for white tulips and appears to come true from seed, though the variegation is scarcely apparent until flowering time.

White reflections

Pools, streams and wet areas are a great asset to any garden, giving scope for plants which are only happy in sodden ground and for effects of cool lushness that can only be achieved with that unique combination of dark rich mud and clear water. White flowers are particularly lovely in an aquatic setting, either in contrast to the abundance of luxuriant foliage or adding their own gleam to that of the water. Among true aquatics, pride of place goes to the white water lilies – *Nymphaea alba*, the scented *N. odorata* and their many hybrids. Largest of all is 'Gladstoniana', whose multi-petalled saucers reach 10in (25cm) across. For small pools, or even tubs and sinks, choose *N. odorata* 'Minor' or the free-flowering *N. candida*. Whenever I see a flotilla of water lilies, I am reminded of Laurie Lee's description of how 'they poured from their leaves like candle fat, ran molten, then cooled on the water'. Second only to luxuriating in the scents of white flowers at twilight is gazing at white water lilies as they float among reflections of sky and clouds – a fascination which absorbed Claude Monet for the last years of his life in the garden he created at Giverny.

Less familiar is water hawthorn (*Aponogeton distachyos*), whose long oval leaves sit flat on the surface, interrupted by forked clusters of heavily scented white flowers which turn green as they fade. It often flowers from early summer until winter and makes an unusual and delightful, if sometimes rather inaccessible, cut flower. Also worth a place is the bog arum (*Calla palustris*), with shiny heart-shaped leaves and white spathes in summer; the Japanese arrowhead (*Sagittaria sagittifolia* ssp. *leucopetala*) and its double form 'Flore Pleno', which have attractive pointed leaves and stock-like flowers in late summer as the leaves start to yellow; and bogbean (*Menyanthes trifoliata*), a relative of the gentians, whose exquisitely fringed white flowers repay close examination.

Two of the most striking plants for wet situations are aroids: the South African arum or calla lily (*Zantedeschia aethiopica*) which is more reliably hardy under water than in the open ground if planted at least 6in (15cm) beneath the surface; and

the magnificent east-Asian skunk cabbage (*Lysichiton camtschatcensis*), whose imposing white spathes appear in spring just before the vast cabbage-like leaves start to unfurl. It begins flowering slightly later than its cousin, *L. americanus* – the same thing in primary yellow. If grown together the occasional hybrid with cream flowers will be produced but is not a patch on its brilliant parents. Flowering at the same time as skunk cabbage is the familiar marsh marigold. Its white form, *Caltha palustris* 'Alba', is a Himalayan variant. Though producing fewer flowers than the usual yellow species, it often has a second flush in autumn.

Irises, with their sword-shaped leaves and graceful flowers, are indispensable near water. *Iris sibirica* 'Snow Queen' and *I. laevigata* 'Snowdrift' are among the best, the former making an excellent bog plant and the latter being happy in shallow water. *Iris spuria* 'Alba' is lovely too, though a very bluish white. Astilbes are equally important, their divided leaves and froth of flowers making a fine contrast to the simpler shapes of the aroids and irises. *Astilbe* x *arendsii* 'White Gloria' and 'Irrlicht' are both good whites, recently eclipsed by the even whiter 'Snowdrift'. White astilbes are also effective planted in masses next to other varieties such as the deep crimson 'Fanal'.

White at night
Sitting in the garden on a warm summer evening at dusk is a favourite pastime for many gardeners. Anyone who has enjoyed this experience will have noticed how colours are first ruddied by the setting sun and then slowly merge into various degrees of greyness as the light fades – all except white flowers, that is, which take on an almost lunar glow. And accompanying these changes from multicolour to monochrome is a general quietening down of the daytime bustle until an almost breathless stillness prevails, with only the last warning call of a blackbird, the first owl's hoot or occasional flitting moth to ruffle the peace. This blissful time scarcely needs improvement but there is one thing (in addition to the chosen drink) which makes the whole thing even more of an indulgence after a hard day, and that is to take a deep breath and inhale wafts of perfume from the becalmed flowers.

Although the scents of many flowers may linger on into the night, planting white evening-scented flowers near seats and paths, and close to windows and doorways will make twilight in the garden even more magical. Best and easiest for this purpose is the tobacco plant (*Nicotiana affinis*), which has sticky long-tubed pure white flowers that are lime green on the outside. In common with many moth-pollinated species, the flowers close during the day and switch off their perfume, opening as night falls and releasing a powerful fragrance.

There are both dwarf and variously coloured hybrids now available which pride themselves in staying open all day, but I still prefer the more untidy superscented original species. I even like the way its flowers screw up against the glare of the sun, to be opened by the magic wand of dusk. The original tobacco plant can reach 36 in (90 cm) and needs to be positioned well back from paths. It tolerates a wide range of conditions in sun or partial shade and in mild areas may even survive the winter. Its performance during the 1976 drought in the British Isles was memorable, outliving and outflowering all else, and though I cursed the searing heat and dearth of rain for what it did to the rest of the garden, I shall be eternally grateful for the experience of sleeping outside on the patio in the scented air of the tobacco plants. Though unsurpassable for fragrance, in appearance it is rivalled by *N. sylvestris*, a more shapely relative whose flowers are longer and more slender, and are borne in one-sided bunches above great bright green lyre-shaped leaves.

Once very popular but seldom seen today is the marvel of Peru or four o'clock flower (*Mirabilis jalapa*), a tender tuberous perennial which can be treated in the same way as a dahlia. It forms a neat bush about 24 in (60 cm) tall with trumpet-shaped flowers which close in the day and open as the light dims. Unfortunately, it is usually only available as seed which produces plants with flowers in various colours – pale pink, cerise, yellow, white or mottled and striped – and the only way to get white-flowered plants is to sow a whole packet the first year, see what comes up and then save tubers of the best. In former times the marvel of Peru was grown in pots solely to decorate rooms used for evening entertaining.

The white form of dame's violet or sweet rocket (*Hesperis matronalis* 'Alba') is one of the most delightful night-scented plants (the name *Hesperis* means 'evening'). The simple four-petalled flowers are nothing special in themselves, but a mass of flowers is exceptionally pretty, perhaps because of their simplicity. One of the most effective plantings I have seen was at Hidcote Manor where a huge drift of it, together with the mauve form, floated amongst tall ferns. Rare and well worth acquiring is the double white form 'Alba Plena', which is even more strongly scented. It does not come true from seed and has to be propagated from any sideshoots which may be produced at the base of the plant. This, together with the fact that it is biennial and has to be propagated regularly, accounts for its rarity.

If the garden seat is situated near a fence or wall, the evening's fragrance can be intensified by garlands of the pale fragrant flowers of honeysuckle, or woodbine as it used to be called. Most honeysuckles have pink-flushed yellow flowers but paler varieties more like the wild one are also available. *Lonicera periclymenum* 'Graham Thomas' and *L. caprifolium* 'Cornish Cream' both have cream flowers which show up well at night. Though not noctural, the scents of other climbers such as jasmine (*Jasminum officinale*), *Clematis flammula* or, in warm areas, *Trachelospermum jasminoides* will linger into the evening, especially if trained into arbours and bowers which reduce air movement.

Night and day, lilies reign supreme in the summer garden. The hardiest species may be planted in the ground where conditions are suitable, but most benefit from the extra attention they receive in pots. They are invaluable for providing a succession of subjects for strategic positions and, though not strictly night-scented, they have an unearthly beauty when veiled in darkness and their pervasive fragrance can still be enjoyed as evening falls. *Lilium regale* is perhaps best of all, a remarkably tolerant garden plant which is equally good-natured, and earlier flowering, in pots. *L. longiflorum* can be first on the scene in early June. Though not hardy it may be gently forced under glass until danger of frosts is past. *L. speciosum* 'Album' has a sweet, gentle scent rather like baby lotion, and its pure white recurved blooms are a bonus in late summer when everything else seems to be fading fast. Last of all, and perhaps most majestic, is *L. auratum*, whose golden-rayed flowers see the first colour changes and mists of autumn.

Another favourite Victorian pot plant was the moth-pollinated tuberose, 'the sweetest flower for scent that blows', according to the poet Shelley. Its scientific name, *Polianthes tuberosa*, means tuberous (not tuberose!) grey flower. It is not grey, of course, but white, and a relative of the narcissus, not a rose.

In Victorian times, tubers were planted at intervals to give flowers all year round. There were double African, American and Italian cultivars, as well as 'The Pearl', a rather shorter variety which is the only one now available. From autumn to spring they were enjoyed in pots, and in summer were added to the border. Being rather ungainly plants with a basal rosette of long linear leaves and a flowering stem that seems as if it will go on forever before the buds develop (it can reach 4 ft (1.2 m)), putting them outside as flowering time approaches seems a good idea. In any case, their fragrance verges on the overpowering – an asset in the twilight garden but somewhat excessive if under your nose in the conservatory. Tuberoses should be started off as early as possible in the year or they will flower too late for you to enjoy outdoors. Plant one to a 6 in (15 cm) or three to a 12 in (30 cm) pot in rich compost and give ample warmth and very little water until growth is under way. They are painfully slow at first but well worth waiting for.

Dicentra spectabilis 'Alba' – the white bleeding heart, a befittingly sensitive plant with delicate foliage and graceful wands of jewel-like flowers.

Gentiana asclepiadea 'Alba' – the white willow gentian, a graceful summer-flowering plant for cool shade.

Lunaria annua 'Variegata' – some white-flowered strains of the variegated honesty come remarkably true from seed, with only the occasional purple-flowered plant among them. The variegation may be scarcely apparent in seedlings but increases dramatically in the run-up to flowering.

ABOVE LEFT Menyanthes trifoliata *(bogbean) – a relative of the gentians, whose exquisitely fringed flowers repay close examination.*

ABOVE RIGHT Iris sibirica *'Snow Queen' – a late-flowering cultivar and one of the finest white irises for moist rich soil.*

RIGHT Nicotiana sylvestris – *an imposing tobacco plant for large borders, with lyre-shaped leaves and a candelabra of scented flowers in later summer.*

White Flowers for the Conservatory

Conservatories and garden rooms bring a whole new meaning to the word 'houseplant'. Both the range and numbers that can be grown by the enthusiast increase enormously, and the fact that they are growing in an extension of the living room, and not in a greenhouse at the bottom of the garden, means that they can be enjoyed at all times of the day and evening, even in winter. Needless to say, conservatories, as well as the tastes and budgets of their owners, vary enormously. I write only from my own experience of one measuring 8ft (2.4m) by 10ft (3m) which is kept at a winter minimum of 45°F (8°C). It is situated so that it gets the sun from late morning until sunset and opens into a lounge which in winter is heated at 65–70°F (17–21°C). Often in winter it is too cold for the sliding door between the lounge and conservatory to be opened but the plants can still be enjoyed through the glass, which extends from floor to ceiling and almost the full width of the wall.

Having a greenhouse as part of your living space gives scope for creating new and exciting surroundings. Regardless of the kinds of plants, the aim should be to create a year-round display that can be enjoyed from whichever room it adjoins as well as inside the conservatory itself. It can have a relaxed Mediterranean atmosphere with citruses, oleanders, bougain-villeas and myrtle, or a more disciplined collection of cacti and succulents – or even, with sufficient summer shading, be transformed into a green and leafy Victorian fernery. Or it can be planned along the same lines as a border, with certain colours of flowers and foliage predominating. White flowers in summer give a cool, sophisticated feeling and are particularly lovely on summer evenings. It is also a seasonally appropriate colour for Easter and Christmas, either alone or combined with Easter yellows and Christmas reds.

White flowers for the conservatory in autumn and winter

If you are a keen gardener in a temperate climate, autumn, for all its glories, is rather a mournful time and winter can be downright dismal. A conservatory can change all that, with tender summer-flowering things carrying on well into autumn and the first spring bulbs and coolhouse pot plants in flower from the beginning of winter. True, there are many lovely plants that flower out in the garden during the harshest months but they seldom reach their full potential as hard frosts, heavy rain, strong winds and the occasional covering of snow take their toll. If bad weather strikes at flowering time, I venture out and carefully dig up small clumps of violets, primroses and suchlike, and put them in pots. In the haven of the conservatory they open fully and release their delicate scent, suffering no ill-effects when replanted. But these are guest appearances: the main cast being composed of tender species which never experience the rigours of the great outdoors, together with a variety of hardy or near hardy bulbs that, thanks to the ingenuity of the horticultural industry, are 'prepared' to bloom out of season indoors.

Long-life bedding plants

Filling a conservatory with flowers in the autumn and winter takes rather more planning than for other seasons. It calls for a mixture of cool-growing houseplants, half-hardy annuals and forced bulbs, some of which can be bought in flower but the remainder needing to be sown or planted with some fore-thought. For instance, most of the half-hardy annuals that one buys in garden centres for summer bedding are sown under glass during the winter to be ready for sale by late spring. Some are even in flower then. Not surprisingly, most are past their best by late summer. However, if you raise them yourself, sowing can be staggered from late winter to early summer so that the latest will continue to flower in the conservatory long after those in the garden have been killed by frost. At first it may seem rather odd to have what are thought of as summer bedding still going strong through the autumn and into winter, but although they are generally treated as half-hardy annuals, many are in fact perennials if kept frost-free. Even if it is not worth keeping them from year to year, they will certainly flower for longer in a cool conservatory.

Lobelia erinus is a good example. If you can rid yourself of images of lobelia as a summer bedding plant in alternate blobs with alyssum or as an underskirt for hanging baskets of geraniums and fuchsias, you will find it a delightful subject for pots in its own right, or as an accompaniment to winter-flowering houseplants such as cyclamen and chrysanthemums. Being a perennial in its native southern Africa, it needs only protecting from frost for a long life and a floriferous one. Its cultivars are beyond reproach as conservatory plants, flowering well into the winter from a late spring sowing and being exemplary in habit – that is to say, they are virtually disease- and pest-free, they flourish in small pots (unlike many other- wise beautiful plants which fall prey to every affliction and constantly need to be potted on if they are to give their best), and their dead flowers disappear as if by magic without dropping or casting petals over all and sundry. There are pure white cultivars available, such as the compact 'Snowball' and trailing 'White Cascade', but they will always produce a few blue-flowered plants which should be carefully tweaked out when they first show colour.

Petunias are shining examples too, with no problem of petal drop, though dead flowers should be removed regularly in winter to prevent mould. The choice is basically between the large-flowered (grandiflora) hybrids, which have luxurious wavy flowers 3½ in (9 cm) across, and the more matter-of-fact medium-flowered (multiflora) hybrids, with basic trumpet-shaped blooms – unless you like the outrageously fancy double petunias. All are available in white, the only real difference being that some shade to yellow in the centre whereas others are green-throated.

Similarly perennial and long flowering if kept frost-free are busy lizzies (*Impatiens sultanii* hybrids). The modern white F1 hybrid cultivars are quick from seed and exceptionally good, with little but habit to choose between them: compact varieties such as 'Super Elfin' being suitable for windowsills and plant benches, and those with a more spreading habit – 'Futura', for example – better for higher-level displays on shelves and in wall pots or hanging baskets. The only trouble with busy lizzies is that, like geraniums (which will also flower through most of the winter, given half a chance), they insistently remind you of their floriferousness with a scattering of fallen petals.

Pansies can be in flower every day of the year but their little faces are never more welcome than at the approach of winter. Towards the end of the summer, trays of winter-flowering varieties are for sale at garden centres alongside sweet williams and wallflowers for spring bedding. If they are potted up separately they will make a cheerful show during even the greyest coldest days. Alternatively, seed can be sown in early summer for winter flowering, with another batch later on in summer to prolong the display into spring. There is a wide choice of plain colours, including cream, ivory and pure white, mostly with a yellow eye, and some – such as 'Floral Dance' – with contrasting navy blue whiskers.

Forced bulbs

Many different bulbs, hardy, half-hardy and tender, can be induced to flower in late autumn and winter given correct treatment. Some will already have undergone special condi- tions when you buy them. They are usually labelled 'specially prepared' and cost slightly more. The secret of success with forcing bulbs is: early planting in late summer; about eight weeks in a cool dark place (such as a shed or garage) for adequate root development; then subdued light and modest warmth – 50°F (10°C) at first, gradually increasing to a max- imum of 65°F (18°C) and full light after another week or so. Forcing is the art of gentle persuasion, not brute force; nothing is to be gained – indeed, everything will be lost – by attempts to rush the process. There is nothing sadder than a bowlful of bulbs distortedly trying to flower at soil level after being given an overdose of light and warmth. Although usually regarded as annuals, many of the hardy ones can be planted outdoors after flowering to add to the garden display in subsequent years.

Hyacinths (*Hyacinthus orientalis* hybrids) are the best known subjects for early forcing, either in pots of compost or the specially shaped hyacinth jars filled with water. The best single white is a variety grown since Victorian times: 'L'Inno- cence'. 'Carnegie' is larger and later. Multiflowering hyacinths are rather more expensive but make a more graceful, informal display. Having more stems of fewer, smaller flowers, they show little tendency to topple over just when they reach their best, as do the large hyacinths. They can be treated in the same way as the large-flowered varieties but flower later. 'Snow Princess' is the one most usually offered.

With hyacinths and narcissi flowering together, the conserva- tory in winter will be as scented as the tropics. The quickest and easiest narcissus for forcing is the *Narcissus tazetta* var. *papyraceus*, known as 'Paperwhite Grandiflora'. It is the only variety which can be put straight into the lounge or conserva- tory after planting (the others need the usual cool spell for adequate root development). However, it is also one of the few that is not worth planting out in the garden after flowering as it is only hardy in very mild areas. The clusters of sweetly scented pure white flowers are produced about six weeks from planting in either pots or in bowls of pebbles and water. If you buy some as soon as they come on to the market in late summer and pot them up at intervals, you will have flowers from mid-autumn all through the winter. For the pebble-and-water method to be successful, a deep bowl is essential. There should be a good 2 in (5 cm) of pebbles, charcoal and rainwater on the bottom, followed by the bulbs surrounded by another 2 in (5 cm) of pebbles to keep them upright. The water level should not go above the bases of the bulbs. Other good

Euphorbia pulcherrima
'Alba' – white poinsettias
complement rooms with
pastel or neutral colour
schemes and combine well
with variegated scented
geraniums in the
conservatory.

Convallaria majalis – *sweet-scented lilies of the valley were great favourites for forcing in Victorian times. The gold-striped form ('Variegata') has the added attraction of decorative foliage.*

bunch-flowered varieties for forcing are 'Cragford', which has clusters of scarlet-cupped white flowers; 'Bridal Crown', with white petals and double yellow and white centres; and 'Cheerfulness', which is similar to 'Bridal Crown' but later. The Chinese New Year narcissus is also worth trying.

Tulips are on the whole less exciting for forcing, with a limited range of varieties – mostly the stocky double ones – and nothing much in the way of scent. 'Snow Queen' is one of the most reliable white forms. They need a cool dark phase and do better if the brown skin is removed before planting. Almost any other bulbous plant – snowdrops, crocus, dwarf narcissi and iris, grape hyacinths, scillas – can be flowered early in the conservatory by planting in pots and leaving in a sheltered place outdoors or in a cold frame until the flower buds are just showing colour. Early planting (in late summer), some protection from frost while making roots and shoots, together with the gentle warmth of a cool conservatory when they are close

to opening, will ensure that they flower several weeks before those in the garden. They last no time at all in the warmth of the living room but in a cool conservatory will give a display that justifies all the effort and expense.

Some bulbous subjects need rather different treatment for successful forcing. For guaranteed high-speed flowering the giant *Hippeastrum* hybrids (sold as amaryllis) must be kept at a constant 65°F (18°C). With this amount of warmth they will flower within six to eight weeks. Transferring them to the conservatory as the buds open will prolong the life of their spectacular blooms. There are a number of white varieties – all excellent – but the only one with any significant difference is 'Picotee', which has a fine red line round each petal. It is worth buying the largest bulbs 12–14 in (30–34 cm) you can find as they will produce two or even three stems of flowers, giving weeks of exotic magnificence in the dead of winter.

Freesias are rather more tricky but worth the effort. They should be planted in sandy compost at the end of the summer and left outdoors until each corm has seven leaves, then brought into the conservatory to flower in late winter or early spring. Staking at an early stage is recommended (fine birch twigs are ideal) as the growths are rather floppy.

A great favourite for forcing in Victorian times was the lilly-of-the-valley (*Convallaria majalis*). They were planted in cones of moss, especially for scenting entrance halls at Christmas, large quantities being imported from Germany for this purpose. It is best to buy large specially grown crowns, though garden plants can be used if the bed has been well fed with manure or compost. Crowns should be started into growth in the autumn in a compost of peat, sand and leafmould. They must have constant moisture and humidity and initially should be kept in the dark at 50°F (10°C) until the shoots are about 4 in (10 cm) high. Then they can be brought into the house, though they will be happier in the lower temperatures and higher humidity of the conservatory or the chill of an unheated entrance hall.

Orchids

Orchids have a reputation of being difficult to grow and almost impossible to flower. For the majority of species such warnings are justified, but there are a number of modern hybrids, and even a few species, which are reasonably adaptable. Most flower between autumn and spring and last up to two months in full glory so are worth the trouble – and the initial expense. Being epiphytes (plants which perch up in trees rather than growing in the ground), their requirements are rather different from the general run of houseplants. They need: a cool sunny rest with very little water in winter when 'in spike' and flowering; plenty of warmth, moisture, shade and humidity in summer; and less of everything as growth begins in spring and finishes in autumn. Feeding should be dilute, with a high potash mixture given towards the end of summer.

A conservatory kept at a winter minimum of 45°F (8°C) is suitable for coolhouse orchids – those from higher altitudes of the tropics. Cymbidiums, both full-sized and 'mini-cymbids', come into this class. Specialist orchid nurseries carry the best selection but a few can be found at almost any garden centre. Pure whites are hard to come by, though Australian breeders have produced several which have only the palest cream spotting on the lip. More readily available are whites such as 'Pearl Balkis', with crimson-marked lips. Plants of the same hybrid may vary in colour so it is much the best to choose one when in flower: the mini 'Showgirl' can be anything from blush-white to pink.

The easiest cool-growing white-flowered orchid species is *Coelogyne cristata*. It grows wild at quite high elevations in the Himalayas and, apart from the cool dry rest in winter, is undemanding in its requirements. It can be grown in a shallow pot or hanging basket filled with orchid compost and top-dressed with sphagnum moss. Repotting should be a rare event, done only when the plant has crept well over the edges of its container. During the rest period it is normal for the pseudobulbs to shrivel, plumping up again when growth and watering get underway in the spring. For years I grew and flowered one on a kitchen windowsill in a house with no central heating. It was therefore frost-free (just!) in winter and had no more than average British summertime temperatures, and still it flourished, generously responding to such harsh treatment with arching spikes of large crystalline white flowers every New Year. There is a pure white form and one with lemon-yellow on the lips but the species, with its egg-yolk yellow markings, is perfectly lovely. Unfortunately, like cymbidiums, it has no scent.

Some of the modern *Phalaenopsis* hybrids, popularly known as moth orchids, are also reasonably successful as houseplants but need more warmth than cymbidiums. The cool conservatory is all right for summer but the bathroom or kitchen is the best place from autumn to spring. Ideally they should have a minimum temperature of 70°F (21°C) in the day, dropping to 60°F (16°C) at night, with even higher day temperatures in summer. Mine cope with cool conservatory conditions in the summer, with the night temperature falling as low as 50°F (10°C), but are cosseted indoors during the winter. Moth orchids do not have pseudobulbs in which to store water so are not equipped for long dry periods. They must therefore be kept moist all year round, though like all epiphytic orchids need a free-draining bark-based compost so that the roots have access to both air and water. A moth orchid in full flower is so lovely that it is worth a little extra trouble to get things just right. The arching sprays may be over 24 in (60 cm) long, with twenty or so long-lasting butterfly-like flowers 3 in (7.5 cm) across. They come in yellows, pinks, white and various patterns but the pure white ones with a touch of colour on the lip beggar all description. When the main flower spike has finished, it should be cut back to where the first flower was produced and not right back to the base. If the plant is strong and healthy (i.e. if each new leaf is as large or larger than the last), it will often sprout a new flower spike (or more than one) from that point and be almost continuously flowering.

Winter-flowering pot plants

There is a surprisingly good range of flowering pot plants in garden centres during the winter, some of which will last in good condition for months if kept in cool conservatory conditions. The dwarf *Cyclamen persicum* hybrids come in many colours, including pure white and pink-nosed white, and many are scented. Splendid plants can be bought quite cheaply, but they can be grown from seed and should flower the first winter after sowing. The familiar large cyclamen (*C. persicum* 'Giganteum') can also be grown from seed, the easiest varieties being the new F1 hybrids which flower within seven or eight months.

The green-fingered might also like to try winter-flowering primulas from seed. *Primula malacoides*, the fairy primrose, should be sown in late winter or early spring for winter flowering and grown in cool moist conditions. There are several good whites, truly fairy-like in their effect, including the double 'Snow Storm' and yellow-eyed 'Nordlicht'. *P. obconica* is larger all round and nowhere near as charming but makes a worthy long-lasting winter pot plant. Again, seed should be sown early for strong plants. The first F1 hybrid white obconica, 'Juno White', is impressive, with larger flowers, a more compact habit and an earlier and longer flowering period. Both of these primulas are perennial but are seldom worth keeping for a second year.

In recent years white poinsettias (*Euphorbia pulcherrima*) have become increasingly popular for Christmas. Their creamy-white bracts look well in certain colour settings or can be used alongside the red or pink forms to provide a striking contrast. They should really have a winter minimum of 55°F (13°C) but do not seem to mind the few weeks of Christmas and New Year at the cooler temperatures of the conservatory.

Perhaps the loveliest winter-flowering pot plant is *Jasminum polyanthum*, a Chinese jasmine with dark semi-evergreen leaflets and panicles of delightfully fragrant white flowers. Plants flower the first year from cuttings and small specimens can be bought quite cheaply and grown on year after year. It is not hardy but needs only a winter minimum of 41°F (5°C) to survive, though it will flower earlier if the thermometer creeps into the 50s (over 10°C). The flowers are long lasting in cool conditions and are produced right through the winter and early spring. For successful flowering, give the plant a sunny position in the summer to ripen the wood. Prune after flowering, repot, and train the new growths as necessary. (Do not be tempted to cut back the new growths, as this is where the next season's flowers will be produced.) A high potash feed, specially formulated for flowering plants, may be given towards the end of the summer to encourage flowering.

White flowers for the conservatory in spring

Most of the plants mentioned under winter will go on flowering into the spring, especially if the display of bulbs is staggered by leaving some pots longer in their cool plunge or outdoor stage. I always leave some pots outside on a plant bench until they are almost in flower. They are therefore not forced but can be enjoyed to the full at their normal flowering time, come hail or shine. Crocus, tulips, irises and fritillaries are my favourites. Specially recommended is the *Crocus chrysanthus* cultivar 'Snow Bunting', which produces numerous pure white flowers with a deep yellow throat in early spring, and a hybrid of the water lily tulip (*Tulipa kaufmanniana*) called 'Ancilla', a colourful variety which is pink outside and white inside with a red-rimmed yellow centre. Both crocus and tulips open wide in the warmth of the conservatory, a glorious sight which may be missed some years in the garden. The delicate *Iris reticulata* can do well outdoors but some years is nibbled by slugs or beaten down by the elements. As an insurance against missing this fragile flower and its evanescent fragrance, I always plant newly acquired bulbs in pots, leaving them to their fate in subsequent years. There is no alba form as such, but 'Natasha' is a lovely thing with bluish-white flowers, dashed orange on the falls. The great advantage of growing fritillaries in the conservatory is that they can be positioned so that their beautifully chequered interiors can be appreciated. White forms of the snakeshead fritillary *Fritillaria meleagris* are both fascinating and lovely, either on their own or alongside the usual dusky purple form.

Early annuals

We are accustomed to sowing seeds of annuals outdoors in the spring, but much more vigorous plants can be produced by starting them off in late summer or early autumn in the cool conservatory and growing them on in pots. Generally speaking, annuals are opportunist species which expend all their energy in a rapid burst of flowering and setting seed as soon as conditions are favourable. If sown at the end of one growing season, rather than at the beginning of the next, they have a long wait and bide their time by producing a more extensive root system and larger leaves – all the better to fuel an even bigger orgy of flowering when warmth and the time comes. Pots of these magnificent over-wintered annuals make their summertime efforts in the garden look feeble in comparison, and provide a spectacular display for the conservatory in the spring. Love-in-the-mist (*Nigella damascena*) can be highly recommended for this treatment. The pale denim blue 'Miss

Phalaenopsis *hybrid – moth orchids are southeast Asian in origin. The modern hybrids do well as houseplants, given sufficient warmth and humidity.*

White flowers for the conservatory in summer – to the left: Lilium longiflorum *'Gelria'. Upper shelf:* Campanula isophylla *'Stella White' (still in bud);* Saintpaulia *hybrid (African violet);* Pelargonium peltatum *'L'Elégante';* Eustoma grandiflorum *(palest pink form). Lower shelf, back row:* Fuchsia *'Ting-a-ling';* Zantedeschia rehmannii; Spathiphyllum wallisii. *Lower shelf, front row:* Sinningia *(gloxinia) 'Mont Blanc';* Exacum affine *'Rosendal White' (Persian violet); white petunias;* Streptocarpus *'Albatross'.*

Jekyll' and clean white 'Miss Jekyll Alba' look delightful together, their delicate flowers and ferny foliage making a good foil for chunkier plants such as the heavily scented pastel-coloured Brompton and East Lothian stocks (*Matthiola incana* hybrids). Rather similar but less graceful are cornflowers (*Centaurea cyanea*), which come in a range of blues, pinks, purples and white, the best white being 'Snowman'. Another slender upright annual is larkspur (*Delphinium ajacis*). The varieties vary in height from 1–4 ft (30 cm–1.2 m), and again come in all shades of the blue-pink spectrum. At the top of the range is 'Imperial White King' with 3–4 ft (90 cm–1.2 m) branched spires of white mini-delphiniums.

Quite different in appearance is the bushy *Godetia grandiflora*, a Californian annual that also does well from a late summer sowing. The compact varieties reach 12–15 in (30–38 cm) and make showy plants whose foliage is scarcely visible beneath the masses of large azalea-like flowers. They come in every conceivable shade of pink, red, salmon and mauve, together with bicolours, both single and double, and it is with a certain relief that the eye alights on a white variety such as the charming 'Duchess of Albany'. Godetias flower well in 5 in (12.5 cm) pots and John Innes No. 1. They tend to produce leaves at the expense of flowers if too well fed.

Windowsill orchids

Spring sees the flowering of the so-called windowsill orchid, *Pleione*. It is rather a misnomer, as they are unlikely to succeed after the first year in such a position, and one of the great beauties of pleiones is that, given the right treatment, they will increase from year to year until you have large pots full of them. A mass of pleiones in flower is a captivating sight, as they bloom on short stalks before the new leaves emerge so that the whole of the surface is covered in disproportionately large and perfect orchids with no visible means of support. *P. formosana* 'Alba' is one of the finest and easiest. The green pseudobulbs should be half-buried in orchid compost and given shade and plenty of moisture (rainwater only) in the growing season. When the leaves start to yellow in autumn, watering should stop. The pseudobulbs can be removed from the compost and kept dry until repotting the following spring. They must be handled very carefully as the flower buds appear as shoots on the side of the pseudobulbs and are easily damaged. Pleiones can be stood outside during the summer or kept year-round in the conservatory. They will be happy with a winter minimum of 45°F (7°C). Once you have mastered pleiones, try calanthes. Similarly deciduous, but with silver pseudobulbs and tall multi-flowered spikes, they need more warmth, richer compost and greener fingers!

Lilies for Easter

The Madonna lily (*Lilium candidum*) was once the traditional Easter lily, but being unpredictable it has now been superseded by *L. longiflorum*. This magnificent species is ideal for the conservatory and can be brought into flower at almost any time of the year. Dormant bulbs are available in late summer and early autumn. For spring flowering they should be planted promptly and kept outdoors initially so that the roots are well developed before the shoots begin to grow. *L. longiflorum* is a stem-rooting species, which means that the bulbs should be positioned low down in the pot as it roots more from the stem than from the base of the bulb. (It is important to know whether a lily is a stem or basal rooter for successful cultivation. If stem rooters are planted just below the surface they will have no room to make roots; conversely, if basal rooters such as *L. martagon* are planted deep in the pot, the stems may rot.) When the weather turns frosty, pots of *L. longiflorum* must be brought into the conservatory as it is not hardy. Once the shoots are well above the surface, the lilies can be forced if necessary by moving indoors to a temperature of 60°F (16°C). A year-round display of *L. longiflorum* can be produced by staggering planting times. For early summer flowering, the bulbs should be potted up in the winter; for flowering in the autumn or winter, the dormant bulbs can be stored in peat at 40–45°F (4–7°C) – in polythene bags in the fridge, for example – and planted in spring or summer.

Other species can also be induced to flower early if kept frost-free after planting in autumn and given increased warmth once the buds are showing. With this regime, *L. martagon* 'Album' will flower in late spring, followed by *L. regale*, *L. speciosum* and *L. auratum* at intervals during the summer. Impressive displays can be obtained by planting three top-size bulbs in a 12 in (30 cm) pot of peat-based compost. *L. martagon* is not noted for its scent, but the others will fill the conservatory, if not the whole house, with perfume. Those who dislike strong scents should grow the Asiatic hybrids instead. The sparkling white 'Mont Blanc' and 'Sterling Star' make excellent pot plants and respond well to forcing, as does the ivory 'White Happiness'. All have upward-facing, widely opening flowers with a sprinkle of maroon dots.

Arum lilies (*Zantedeschia aethiopica*) are not lilies at all, but aroids. The spathe, which encloses the yellow spadix, is pure white, faintly perfumed and has an almost skin-like texture. These sculptural inflorescences are closely associated with Easter and are in great demand for flower arrangements at that time. Imported blooms are extremely expensive but they can be grown quite easily in the conservatory and, with sufficient warmth, may flower in time for Easter. Although they grow from tuberous rhizomes, a dormant period does not come naturally to them. In their native southern Africa, they are evergreen in wet areas and only call upon dormancy as a last resort in seasonally dry regions or during droughts. I dry mine off at the end of the summer for no better reason than they are large greedy plants and it is me that could do with the rest. In the normal course of events, I entice them back into growth at the first signs of spring and they flower in the summer, but earlier repotting followed by the lashings of water and food they enjoy would bring them into bloom for Easter. It's all a question of timing.

There are a number of different arum lilies that merit a place in the conservatory. *Z. aethiopica* 'Crowborough' is the hardiest and most widely stocked but can reach 3 ft (90 cm) tall and almost as much across. Equally large in habit is the unusual variety 'Green Goddess', which has small spadices and huge wide open spathes which are green at the apex. *Z. albo-maculata* cannot be compared in terms of blooms (they are ivory, narrowly funnel-shaped and about half the size) but it is a more manageable plant with fewer, smaller and more handsome sagittate leaves, marked with translucent dots. *Z. rehmannii* is known as the pink arum but many plants have almost white flowers. It reaches only 15 in (38 cm) and has arching lanceolate leaves. Both these species need a decided rest in winter and make good pot plants, even if they are of no value as cut flowers.

Cool whites for the conservatory in summer

Throughout autumn, winter and spring the conservatory enables plants and plant lovers alike to bask in unseasonal warmth, enjoying an abundance of flowers whose perfection is unmarred by cold winds and sudden downpours. When summer comes temperatures are more nearly the same inside and out but the additional protection afforded by the conservatory still gives it the edge over outdoor temperatures, especially at night. For the plants this means near-tropical conditions; for us, a pleasant setting for an evening's relaxation after busy hours of watering and tending the burgeoning vegetation.

When it comes to choosing plants for the conservatory in summer, for once the gardener is spoilt for choice. Yet among the attention-seeking lilies and floribundant geraniums and petunias, are many plants with more subtle charms: smart evergreen shrubs, futuristic daisies, eccentric nocturnal cacti, velvety gesneriads and demure Persian violets . . .

Summer-flowering tender shrubs

The best known summer-flowering shrub is undoubtedly the fuchsia. Creating fountains of bells in hanging baskets or as a pyramid or standard, its colourful swinging bells are compulsory summertime viewing in almost every garden. It is easy to grow and quick to flower from seeds or cuttings started in spring (though named varieties will not come true from seed), so much so that it is often treated as an annual. There are numerous white varieties, ranging from the small and clean-cut which would be just right for dangly earrings, to large doubles that are reminiscent of frilly underwear. My favourite is 'Ting-a-ling' which is somewhere between the two, unfussy but rather too big for earrings.

Citruses have been popular as conservatory plants since Victorian times for their attractive combination of evergreen foliage, white flowers (whose pervasive scent takes you back to Mediterranean holiday haunts), and colourful edible fruits. The calamondin orange (x *Citrofortunella mitis*) is a good choice for a small conservatory, reaching only 18 in (45 cm) and flowering and fruiting when much less than this. It does, however, need more warmth than most of the other species, preferring a minimum of 55°F (13°C). The others get up to 3–4 ft (90 cm–1.2 m) as pot plants. Citrus fruits can take up to a year to mature, with the result that ripe fruits are still on the tree when the next season's flowering takes place. This makes them of year-round interest as conservatory plants, with the promise of the occasional home-grown orange or lemon as a bonus. Serious citrus lovers will buy a named variety from a nursery – those grown from pips are often sterile.

Myrtle (*Myrtus communis*) is another Mediterranean shrub with a great deal to recommend it. Its faultlessly tidy evergreen leaves are delightfully aromatic and have culinary uses in addition to providing a perfect background for the flowers – delicate things which are more stamen than petal and faintly scented. The blossom is often followed by aromatic purplish-black berries which are used in Middle Eastern cooking. There is an attractive variegated form ('Variegata') and a dwarf form, variously known as 'Tarentina', 'Jenny Reitenbach', 'Microphylla', and 'Nana', which has a variegated version too. Myrtle reaches about 3 ft (90 cm) as a pot plant and can be tidied up in the spring to maintain a good shape. The compact form gets to about half the size. Like citruses, myrtles enjoy plenty of warmth in the summer but tolerate a minimum winter temperature of 45°F (7°C).

At the other extreme are sensitive souls like the gardenia. It needs much higher minimum winter temperatures for a start – near 60°F (18°C) – so can only spend the summer in a cool conservatory and must overwinter as a houseplant. Unfortunately, wherever it is kept, it will probably find something not to its liking and promptly begin a campaign of yellowing and dropping leaves and flower buds, which is sure to drive any plant lover to distraction, especially one who has just paid the kind of price commanded by these prima donnas. Part of the problem, I suspect, is that they dislike being moved and therefore, however green-fingered their new owner, may already be suffering from changes in conditions experienced during the marketing process. And, in addition to their temperamental disposition, gardenias are like magnets for all known greenhouse pests and diseases. As one Victorian gardening manual warns: 'Gardenias ... become more infested with insects than is usual with even ordinary occupants of warm houses'. One has to become as vigilant as a monkey with its young, forever inspecting them most minutely for the first sign of infestation. However, those that take up the challenge and succeed will be amply rewarded, for their sophisticated fragrance is beyond compare.

The most commonly grown gardenias are the double forms of the Chinese *Gardenia jasminoides*: 'Florida', which flowers in the summer and is the easiest; and 'Veitchiana', a winter-flowering variety which needs a winter minimum of 60–65°F (16–18°C) to succeed. Gardenias can, with luck and good management, get quite large but in any case flower better when young. If you look closely at the plants for sale in most garden centres (and you should, to make sure pestilence has not yet struck), you will see that they mostly consist of three rooted cuttings to a pot which, in spite of their flower buds, do not look more than a few months old. So, if your gardenia survives, you should celebrate by propagating it the next spring. Lateral shoots, taken with a heel, root quite easily in a mixture of peat and sand at 70°F (21°C) and with sufficient warmth and humidity may flower the first year. Home-grown gardenias should prove a lot easier to grow on than plants which have suffered the slings and arrows of packaging, transportation, and sales outlets.

Addicts of heavily scented white flowers who find gardenias well nigh impossible, may do better with *Stephanotis floribunda*. This tropical twiner from Madagascar has leathery oblong evergreen leaves and clusters of tubular white flowers whose thick waxy texture, large size and pervasive fragrance makes them popular for wedding bouquets. Though needing plenty of everything but full sun in summer, it tolerates higher light levels and temperatures as low as 50°F (10°C) in the winter.

Another magnificent conservatory shrub that should carry a warning is the datura, commonly known as angel's trumpets. A number of different daturas are in cultivation, the most striking white-flowered ones being *Datura cornigera*, its semi-double form 'Knightii', and *D. suaveolens*. For a start, daturas get very large – 8ft (2.4m) or more tall and as much across – and rapidly outgrow their welcome in all but the largest conservatory. Then, as soon as they begin to show the slightest signs of being starved, red spider mite and white fly descend upon them from miles around. (There is also the danger that they can become infected with tomato wilt virus if the gardener smokes tobacco, grows tomatoes and handles the plants – daturas, tobacco and tomatoes being members of the same family and sharing some common ailments.) As an insurance against ill-health, they need rich compost, generous feeding and ample accommodation – the 12in (30cm) pot usually recommended is woefully inadequate. The good news is that they take drastic pruning and need very little warmth or moisture during the winter. (However, though naturally ever-green, they may lose their leaves if kept at a winter minimum of 45°F (7°C). They can also be stood outdoors in a sheltered sunny position in the summer, which minimises pests, or at least prevents them spreading to the rest of the conservatory. Both the species mentioned produce large, soft, light green leaves and huge white pendulous flowers that are beautifully scented. In spite or because of the difficulties, a well-grown datura is the greenhouse gardener's pride and joy.

Easier and no less arresting in its way is the 'Whirligig' version of the sun-loving African daisy *Osteospermum tauranga*. Though not exactly a shrub, it makes a bushy plant 24in (60cm) or more high, with unpleasant-smelling foliage and unusually stylish flowers that never fail to raise a smile – each dazzling white, blue-backed petal being pinched in the middle. Its other attractions are that it is extremely easy to propagate from cuttings, quick growing, tough and tolerant of low temperatures.

Scented summer evenings in the conservatory

The conservatory can be at its loveliest and most hospitable on summer evenings. The heat of the day has diminished but enough is retained to make it a welcome retreat when a chill descends over the garden, and on frankly miserable summer evenings it gives something of the pleasure of being in the garden with none of the discomforts. For these occasions it is worth cultivating some interesting tender plants which bear powerfully fragrant flowers that for the most part open in the evening and close at dawn.

Night-flowering species are found in a wide range of families. One of the most spectacular is a West Indian cactus,

RIGHT Osteospermum tauranga *'Whirlygig' – an easy, quick-growing African daisy whose stylish flowers never fail to raise a smile.*

Selenicereus grandiflorus (syn. *Cereus grandiflorus*) whose common name is the torch thistle or night-flowering cereus. The Greek word *selene* means 'moon' and in Thornton's *Temple of Flora* (1797–1807) there is an evocative mezzotint engraving of the cactus's huge, pale, multi-petalled bloom against a background of the night sky, the moon and a church clock with the hands at midnight. The flowers of this cactus have an arresting beauty and it is worth gathering a small audience to share with you the fascination of watching them open. To quote a Victorian account:

They begin to open between seven and eight o'clock in the evening, and are fully blown by eleven. By three or four o'clock in the morning, they fade; but, during their short existence, there is hardly any flower of greater beauty, or that makes a more magnificent appearance. The calyx of the flower, when open, is nearly 1 ft [30 cm] in diameter; the inside, being of a splendid yellow colour, appears like the rays of a bright star; the outside is of a dark brown. The petals, being of pure white, contribute to the lustre; the vast number of recurved stamens in the centre of the flower make a fine appearance. Add to all this the strong, sweet fragrance, and there is scarcely any plant which so much deserves a place in the stove, as this, especially as it may be trained against the wall, where it will not take up any room.

If the mention of training a cactus against a wall sounds rather odd, it is because in habit the night-flowering cereus is indeed very odd, consisting of a five- or six-angled prickly rope which snakes along, half climbing and half trailing over any obstacle in its path. It should therefore be taken in hand (gingerly, as it is well-armoured) at an early stage and trained out of harm's way, guiding it up a wall or to the ridge and then along, which should keep it busy for several years. With such an unmanageable creature as this, repotting is a painful, if not farcical, procedure and it is best to give it a fair-sized pot to begin with. John Innes No. 2 plus some extra sand or grit should see to its nutritional needs, with ample moisture, weak feeding and warmth in summer but a cool dry rest in winter. The night-flowering cereus is not difficult to grow. It is usually propagated by cuttings (which can be obtained from a specialist cactus nursery) but takes several years to reach flowering size.

The moon flower, *Ipomoea bona-nox* (syn. *Calonyction aculeatum*) is another spellbinder and has the advantage of flowering within a few months from seed. Though perennial if kept at a winter minimum of 55°F (13°C), it is usually treated as a half-hardy annual. The pea-sized seeds germinate rapidly at 75°F (24°C) and grow quickly. I usually plant two or three to a 12 in (30 cm) pot and train the mass of thin twining stems and heart-shaped bindweed leaves against the conservatory wall. The spectacular pure white flowers are 4–6 in (10–15 cm) across and give out a fragrance reminiscent of clean clothes that have been folded between bars of soap. They resemble giant morning glories but are flatter and more disc-like, and have nectar tubes 4 in (10 cm) long. Nothing is more magical than to sit in the evening and watch the fragile discs of the moon flower turn their blank faces to the moon – unless it is to seeing them wild in tropical America where magnificent moths are drawn out of the darkness by their perfume and reflected light.

Less sensational but worth growing for its evening scent is the night jessamine, *Cestrum nocturnum*, an evergreen West Indian shrub belonging to the same family as potatoes and tomatoes, the Solanaceae. It has slender, rather drooping branches, ovate leaves 4–8 in (10–20 cm) long, panicles of small tubular greenish-white flowers, and poisonous white berries. The heavy scent of the flowers is said to be narcotic, which is worth remembering if evenings in the conservatory suddenly start to become more relaxing than usual! It reaches 10 ft (3 m) or more but can be kept within bounds by pruning in the winter. Cestrums are quite quick from seeds or cuttings and need large pots and rich compost. A winter minimum temperature of 45°F (7°C) should be sufficient.

Also West Indian and another member of the Solanaceae family is the lady-of-the-night, *Brunfelsia americana* which bears white jasmine-like flowers which yellow with age and are wonderfully fragrant at night. There are narrow and broad-leaved varieties and one known as *pubescens* which flowers more freely. It reaches 4–8 ft (1.2–2.4 m), needs a winter minimum of 50°F (10°C) and should be pruned after flowering. The tips of new growths can be pinched out when 6 in (15 cm) long to encourage a bushy habit.

Perhaps easiest of all night-scented flowers is the variably coloured marvel of Peru (*Mirabilis jalapa*) which has already been described on page 79. Though usually grown as an annual, it is a tuberous perennial and can be treated in the same way as dahlias. Tubers of plants with white or pastel flowers can therefore be saved for the following year. It is quick and easy from seed, the large black seeds being interesting in themselves. If you cut one open, you will see that it is filled with a white powdery substance which was once, believe it or not, used as a face powder.

The Ethiopian bulb *Gladiolus callianthus* (syn. *Acidanthera bicolor*) is also readily available and virtually foolproof. I have never been able to determine whether it is, strictly speaking, a night-scented species, but its flowers have that typical soapy fragrance in spite of the fact that they remain open round the clock. The nice thing about this lovely species is that its white, maroon-centred upside-down-looking flowers are produced late in the summer when everything else is beginning to look rather jaded.

Gesneriads

The family Gesneriaceae includes many velvety-leaved ornamental species which make excellent subjects for the conservatory in summer. Being relatively small and shade-loving, they are handy for the spaces between large pots of more voluminous plants. Best known is the African violet (*Saintpaulia ionantha*). Hybrids of this species are so floriferous that they bloom several times a year. They enjoy the warmth and humidity of the conservatory in summer but need a temperature of at least 55°F (13°C) in winter, which may be more easily maintained in the house. Plants with pure white flowers are not as common as pinks and purples and getting hold of a good one may take time. I found an absolute beauty once, the only one amongst dozens of the other colours, with large single crystalline white flowers in such numbers that only a ruff of green was visible beneath them.

When you do come across a particularly fine plant, either purchased or grown from seed, it is well worth detaching the odd leaf occasionally to produce a successor with identical characteristics. Leaves with about 2 in (5 cm) of stalk root readily in a mixture of sand and peat at 70°F (21°C). African violets can also be grown from seed in sandy lime-free compost at the same temperature. 'Fairy Tale White' is one of the few named white-flowered varieties currently available as seed.

Also African in origin are the Cape primroses, hybrids derived mainly from *Streptocarpus rexii*. They too are evergreen perennials which enjoy warm, semi-shaded conditions (mine overwinter in the cool conservatory but would probably be happier a few degrees warmer) and, like African violets, can be propagated from leaves (or even sections of leaves) and will flower the first year from seed. The flowers are somewhere between pansies and foxgloves in appearance, borne on wiry stalks above the strap-shaped leaves which have more the texture of car upholstery than velvet. Over the years, hybrids have got bigger, more colourful and more frilly, with heavy contrasting markings in the throat, losing much of the grace and charm of old cultivars such as the violet-blue 'Constant Nymph'. Those who like the modern look can grow 'Lipstick White', a high-speed F1 hybrid, from seed. It has large white flowers, crimped round the edges and rayed with purple in the centre. I will remain faithful to 'Albatross', whose smaller plainer flowers have a touch of lemon yellow in the throat and look like albino versions of 'Constant Nymph'.

The South American branch of the family has given us gloxinias, achimenes and episcias. The most widely grown of these are the gloxinias, hybrids of *Sinningia speciosa*. They need similar growing conditions to African violets and Cape primroses but die back in the autumn to tubers, which must be stored in small polythene bags of peat at room temperature. In late winter the dormant tubers can be started into growth again by dampening the peat and emptying it into a seed tray. If the tubers are nestled into this and placed in a propagator or airing cupboard at 70°F (21°C), they will soon show signs of life. When the new shoots are an inch (2.5 cm) or so high, the tubers should be potted up separately in a peat-based compost, positioning them so that the tops of the tubers are just level with the surface. When the foliage is well developed the temperature should be lowered slightly and by early summer, when buds are showing, the plants can be transferred to the cool conservatory. If constant high temperatures can be maintained, gloxinias may be raised from seed and flowered the first year. 'Mont Blanc' has huge, pure white, upturned bells with frilled margins. There is also a variety called 'Tigrina', whose flowers are basically white, heavily dotted with red or purple.

Achimenes or hot-water plants were at their height of popularity in the nineteenth century and have everything to recommend them as summer-flowering conservatory plants, apart from the fact that they drop their open flowers if moved about (which may be why they have never caught on commercially). They grow from funny little scaly rhizomes, which at first glance look like catkins or, worse still, caterpillars or grubs, and can give you quite a surprise if you tip out a dead plant and are not prepared for the sight of them in the compost. All they need is to be dotted around about 1 in (2.5 cm) deep in peat-based compost, at the rate of about eight to a 6 in (15 cm) pot and kept warm and just damp. Even a single tubercle makes a fair show in a 3 in (7.5 cm) pot. They take ages to come up but soon make up for lost time, producing bushy semi-erect or semi-pendant plants, depending on which way you look at it, and a profusion of pansy-like flowers all summer long. The tubercles meanwhile are undergoing exponential growth and by the time they finally give in to dormancy, you will have more vegetable caterpillars than you know what to do with.

There is an old wive's tale that you should only water achimenes with hot water, hence their odd name. It is certainly true that they, like most tropical plants, may suffer physiologi-

Gladiolus callianthus (*syn.* Acidanthera bicolor) – *an Ethiopian bulb which blooms in late summer and has the typical scented soap smell of moth-pollinated species, though the flowers remain open all day.*

Achimenes dulcis – *hot water plants, as they are commonly known, are easy and floriferous pot plants for the house or conservatory in summer. This wild species is less often seen than the large-flowered hybrids.*

cal disorders by being watered with cold water straight from the tap, but water at room temperature is quite adequate. Amazingly, some people take their name literally and manage to water the plants by pouring extremely hot water into the saucer without killing them.

Achimenes mostly come in pinks, purples and violets but there are some excellent white ones that are worth seeking out. The prettiest species is *A. dulcis*, which has grey-green downy leaves and pure white flowers which are more trumpet-than pansy-shaped. Of the Victorian varieties, 'Margaretta' (syn. 'Margarita', 'Longiflora Margaritacea') is the best spotless white and 'Ambroise Verschaffelt' (syn. 'Sugar Plum') the finest fancy white with bright purple 'veins' radiating from the centre. More recent whites include 'Moonstone', a floriferous trailer; 'White Giant', an American hybrid with purple-marked blooms; 'White Rajah', a vigorous well-branched large-flowering cultivar; and 'White Knight', a recent introduction and a strong grower with a touch of vivid purple and yellow in the centre.

Connoisseurs of gesneriads will agree that *Episcia di-anthiflora* (syn. *Alsobia dianthiflora*) is the loveliest of white-flowered species. It comes from Mexico and is the smallest of those popular in cultivation. In habit it resembles a strawberry plant, with rosettes which put out runners and develop plant-lets along their length. The dark velvety leaves are only an inch (2.5 cm) or so long – about the same size as the ice-white flowers. As any white flower enthusiast will know, numerous cultivars have names to do with snow ('Snowdrift', 'Summer Snow' and so forth), but here at last is a white flower that actually does bear some resemblance to a snowflake.

Propagating this little gem is easy enough – just detach and pot up a well-formed plantlet – but growing it successfully requires a bit of trial and error. It needs ample warmth and humidity all year round so although it can sojourn in the cool conservatory during the summer, it has to be moved indoors for the winter – ideally to the bathroom or kitchen where the humidity is higher. I grow it in a crocus pot of sandy peat-based compost next to the cooker, starting off with one rosette in each of the side holes and a few on top. It is happy in shade, though probably needs higher light levels to encourage flower-ing, which happens mainly in summer. My plants multiply merrily but are stingy flowerers. Yet in spite of the fact that the flowers tend to be few and far between, each perfect summer snowflake is a treasure worth waiting for.

The Persian violet

The Persian violet is not from Persia, neither is it a violet. Otherwise the name describes it very well, as it is covered with sweet-smelling violet-coloured flowers that look vaguely exo-tic. In fact, it is a gentian from Socotra, an island in the Indian ocean belonging to South Yemen. I have no idea what growing conditions are like in that remote corner of the Middle East, but *Exacum affine*, to give it its scientific name, seems to be a remarkably tolerant little thing providing it is kept above 60°F (16°C). Though it would no doubt prefer constant moisture, I have at times let mine get really dry with no ill-effects. Given a very modest-sized pot of peat-based compost, it makes a tidy bushy plant no more than 12 in (30 cm) high and has the best of all habits for an inhabitant of house or greenhouse in that it never sheds a single leaf or flower or even so much as a petal. When the flowers die, they close up and pretend to be buds for a while before discreetly withering. Further to its credit, nothing but the odd aphid seems to bother it.

Persian violets can be treated as annuals or biennials, flowering within five months from an early spring sowing, or getting that much bushier and flowering earlier if sown in late summer for flowering the following year. The seed is very fine but germinates readily at 65°F (18°C). In recent years, white-flowered plants have come on to the scene but are still by no means common. Seed is however available under the varietal name of 'Rosendal White'. The Rosendal series improves upon what is already a near-perfect pot plant, being even more compact and ball-shaped in habit, very free-flowering and just as fragrant. All in all, white Persian violets must be the prettiest, sweetest scented, most well-behaved and trouble-free plants that you could grow in the conservatory in summer – and they make perfect gifts for lovers of white flowers.

Exacum affine

Plant lists

Hardy Annuals

*Alyssum maritimum** (syn. *Lobularia maritima*) – bushy plant, slightly grey-green foliage, densely covered in rounded heads of tiny honey-scented flowers; self-sows in paths in mild areas; 'Carpet of Snow' – low spreading pools 3–4 in (7.5–10 cm) high and up to 15 in (38 cm) wide; 'Little Dorritt' – compact, more upright plants with pinnacle-shaped flower heads, 6 in (15 cm); 'Minimum' – probably the smallest and particularly good in cracks and crannies, 2 in (5 cm); 'Sweet White' – highly scented, 4 in (10 cm). Well-drained, sunny position.

*Argemone grandiflora** – a Mexican poppy with grey-green, white-veined prickly leaves and crinkled 4–5 in (10–12.5 cm) single flowers, short-lived but plentiful. Average to poor soil, sun; 12–24 in (30–60 cm).

Centaurea cyanea (cornflower) 'Snowman' – long-lasting, pure white cornflowers freely produced. Well-drained, sunny position; 15–24 in (38–60 cm). ✂

Centaurea moschata (sweet sultan) 'The Bride' – an oriental cornflower with narrow grey-green leaves and slender stems of fragrant white flowers which resemble rather elegant dandelions in appearance. Ordinary soil in sun; 16–24 in (40–60 cm). ✂

Chrysanthemum paludosum – a mini-marguerite with feathery grey-green leaves and neat little daisies. Ordinary soil in sun; good in containers; 12 in (30 cm).

Clarkia elegans 'Albatross' – mid-green, ovate leaves and dense spikes of double flowers up to 2 in (5 cm) across. Light, slightly acid soil and sun; 2 ft (60 cm). ✂

Claytonia perfoliata (syn. *Montia perfoliata*) (spring beauty) – a curious little plant with pairs of fresh green leaves fused together round the stem to form a backing for miniature posies of tiny white stars; prefers slightly acid sandy soil in shade; self-sowing and carpeting where happy; 3–6 in (7.5–15 cm).

Crepis rubra 'Snowplume' – a southern European hawksbeard with a rosette of light green leaves and wiry stems of 1 in (2.5 cm) white 'dandelions'. Well-drained, average to poor soil in a sunny position; 12 in (30 cm).

Eschscholzia californica (Californian poppy) 'Milky White'* – ferny blue-green leaves, single silky cream flowers, 3 in (7.5 cm) across, short-lived but plentiful. Poor sandy soil and sun; 12–15 in (30–38 cm).

Godetia grandiflora – compact plant with mid-green leaves and long-lasting funnel-shaped flowers, 2 in (5 cm) across; 'Duchess of Albany' – single, 12–15 in (30–38 cm); 'Double White' – 15–20 in (38–50 cm). Ordinary soil and sun. ✂

Godetia grandiflora

Gomphrena globosa (globe amaranth) 'White' – bushy erect plant with light green, hairy leaves and ovoid everlasting flower heads. Ordinary well-drained soil in sun; 12 in (30 cm). ✂

Gypsophila elegans (baby's breath) – the perfect accompaniment for cut sweet peas; 'Alba Grandiflora' – large white flowers; 'Covent Garden White' – the most widely grown variety, narrow glaucous leaves and much-branched sprays of small white flowers. Ordinary well-drained soil in sun; 18 in (45 cm). ✂

Helianthus annuus (sunflower) 'Italian White' – the nearest yet to a white sunflower, though only 3–4 in (7.5–10 cm) in diameter and definitely cream, shading to yellow around the dark centre; good with bright blue flowers. Ordinary soil in a sunny position, remembering that the flowers will face more or less south; 4 ft (1.2 m).

Helipterum manglesii (syn. *Rhodanthe manglesii*) – a delicate Australian everlasting, available in 'White' and 'Double White'. Poor stony or sandy soil in sun; 12 in (30 cm). ✂

Hibiscus trionum (flower-of-an-hour) – bushy floriferous plant with dark green ovate leaves and short-lived cream, chocolate-centred flowers. Ordinary well-drained soil in sun; 2 ft (60 cm).

Iberis amara (candytuft) – narrow mid-green leaves and neat, densely flowered racemes of small, slightly scented flowers; 'Giant Hyacinth-flowered' – the only readily available white variety (candytuft is generally sold in mixed colours which include white). Ordinary to poor, well-drained soil in sun; 12–15 in (30–38 cm).

Lathyrus odoratus (sweet pea) – tendril climber with pairs of grey-green ovate leaves and (usually) scented flowers. Some of the best whites are: 'Aerospace' – long-stemmed, sweetly scented, pure white; 'Cream Southborne' – vigorous, long-stemmed, with very large cream, scented flowers; 'Jilly' – ivory, classic form but almost scentless; 'Diamond

Wedding' – a scented, long-stemmed, vigorous white; 'Hunters Moon' – deep cream, almost primrose, with a good scent; 'Royal Wedding'* – pure white, beautifully scented large flowers on very long stems; 'Snowdonia Park' – pure white, sweetly scented; 'Swan Lake' – pure white; 'White Leamington' – very large frilly scented flowers on long stems. Remove dead flowers daily to prolong flowering. Slightly alkaline soil, enriched with manure or compost and with a cool, deep root run, in a sunny position; 6 ft (1.8 m) or more. ✕

Lavatera trimestris (mallow) 'Mont Blanc'* – one of the best white annuals; erect bushy plant with mid-green, ovate, lobed leaves and gleaming hibiscus-like flowers; 20–24 in (50–60 cm).

Lobularia maritima see *Alyssum maritimum*.

Malope trifida grandiflora 'Alba'* – a truly beautiful Spanish mallow; slightly lobed, mid-green leaves and hibiscus-like flowers 2–3 in (5–7.5 cm) across with green-starred centres. Any soil in sun; 2½–3 ft (75–90 cm).

Montia perfoliata – see *Claytonia perfoliata*.

Nigella damascena (love-in-a-mist) – 'Miss Jekyll Alba'* – a superb white annual with long-stemmed, semi-double flowers followed by ornamental pods. Well-drained soil in sun; 18 in (45 cm). ✕

Omphalodes linifolia (Venus's navelwort) – grey-blue foliage and numerous small, faintly scented flowers. Ordinary soil in sun or partial shade; 12–18 in (30–45 cm).

Papaver somniferum (opium poppy) 'Paeony Flowered White Cloud'* – waxy grey-green leaves and huge, heavy, double poppies, followed by 'pepperpot' capsules. Ordinary well-drained soil in sun; 2½ ft (75 cm).

Rhodanthe manglesii – see *Helipterum manglesii*.

Salvia horminum 'Clarissa White' – compact plant with broad green-veined white bracts. Fertile well-drained soil in sun; 18 in (45 cm). ✕

Saponaria vaccaria 'Alba' – sprays of dainty flowers, rather like *Gypsophila*. Fertile soil in sun or partial shade; 2 ft (60 cm).

Xeranthemum annuum 'Snowlady' – a wiry-stemmed everlasting daisy. Ordinary soil in sun; 2 ft (60 cm). ✕

Half-hardy Annuals

Abronia fragrans (sand verbena) – a Californian species with an upright habit and heads of night-scented white flowers. Humus-rich sandy soil in sun; perennial if protected from frost; 2 ft (60 cm).

Ageratum houstonianum – neat plants with powderpuff heads of dense fluffy flowers; 'Summer Snow' F1 hybrid – compact white, 6–8 in (15–20 cm); 'White Cushion' – creamy white, 8–10 in (20–25 cm). Sheltered sunny site with moisture-retentive soil.

Ammi majus (Queen Anne's lace) 'Snowflake' – umbels of tiny flowers, similar to cow parsley. Ordinary soil in sun or partial shade; 3 ft (90 cm). ✕

Antirrhinum majus (snapdragon) – the familiar sub-shrubby bedding plant with spikes of tubular round-lipped flowers; 'White Spire' – a vigorous tall sentinel hybrid, 3 ft (90 cm); 'White Wonder'* – pure white, 18 in (45 cm). ✕

Arctotis stoechadifolia (African daisy) – a tall, quick-growing species with narrow toothed grey-green leaves and green-centred pearly white 3 in (7.5 cm) daisies; 2–3 ft (60–90 cm); var. *grandis* – larger all round with blue-centred white, cream or primrose daisies; 3–4 ft (90 cm–1.2 m). Needs well-drained, even dryish soil, and full sun.

Begonia semperflorens (bedding begonia) 'Viva' F1 hybrid* – bright green leaves, pure white flowers. Good soil, sun or shade; 6–8 in (15–20 cm).

Callistephus chinensis (China aster) – long-lasting double asters for bedding and cutting; 'Milady White' – a dwarf bushy wilt-resistant white variety. Ordinary well-drained soil in sun, avoiding the same site in consecutive years, which encourages aster wilt; 10–12 in (25–30 cm). ✕

Chrysanthemum parthenium see *Tanacetum parthenium*.

Cleome spinosa (spider flower) 'Helen Campbell'* – a tall dignified plant with spiny stalks, attractive palmate leaves and spectacular terminal clusters of spidery, long-stamened flowers in late summer. Good soil in sun; 3½ ft (1.05 m).

Cosmos bipinnatus 'Purity'* – elegant Mexican daisies up to 4 in (10 cm) across, with ferny foliage. Light, poor soil in warm, sunny position; 2½ ft (75 cm). ✕

Cosmos bipinnatus *'Purity'*

Datura stramonium (thorn apple; jimson weed) – bushy plant with scented, upright, trumpet flowers followed by large, rounded seed capsules. Poisonous. May self-sow. Ordinary well-drained soil in sun; 2½ ft (75 cm).

Dianthus caryophyllus (carnation; gilliflower) 'White Knight' F1 hybrid – grey-green foliage and long-lasting flowers with strong stems. Perennial but often treated as a half-hardy annual. Well-drained neutral to alkaline soil in sun; 12 in (30 cm). ✕

Dianthus chinensis (Indian pink) 'Snowflake' F1 hybrid – long-lasting, free-flowering scentless single pinks. Well-drained neutral to alkaline soil in sun; 8 in (20 cm).

Eustoma grandiflorum (syn. *Lisianthus russellianus*) (prairie gentian)* – waxy grey foliage and huge long-lasting goblets which stay fresh for a month as a cut flower. Normally violet-blue but F1 hybrid seed includes white and palest creamy pink. Tricky but well worthwhile as a pot plant; prone to mildew and needs careful watering (a little and often in summer, gingerly in winter). Sandy soil with added peat and leafmould; 12–18 in (30–45 cm). ✕

Helichrysum bracteatum (everlasting flower; straw flower) – the familiar 2 in (5 cm) daisy flowers with yellow centres and incurved papery petals; 'Monstrosum White' – large, double, ivory flowers. Flowers for drying should be cut before fully opened and hung in a cool, shady, airy place. Easy in any well-drained soil in sun; up to 4 ft (1.2 m). ✕

Hibiscus moscheutos – a hardy North American perennial, usually treated as a half-hardy annual; serrated ovate leaves and large white to mauve flowers with purple centres. The species is rarely seen; instead you can grow 'Disco Belle White' from seed, which has red-centred flowers the size of dinner plates and flowers in 4–5 months from a late winter sowing. Well-drained soil in full sun; 20 in (50 cm).

Lisianthus russellianus – see *Eustoma grandiflorum*.

Lobelia erinus var. *compacta* (compact lobelia) 'Snowball'* – pure white, but with some blue-tinged plants in every packet of seed. Rich moist soil in sun or partial shade; 6 in (15 cm).

Lobelia erinus var. *pendula* (trailing lobelia) 'White Cascade'* – dazzling white lobelia for hanging baskets. Remove the inevitable odd blue-flowered plants as they begin to show colour. Rich moist soil in sun or partial shade; 12 in (30 cm).

Matthiola incana (stock) 'Dwarf Stockpot White' – small compact plant with heavily scented flowers, blooming in 7–9 weeks from sowing and ideal for pots, 8–10 in (20–25 cm); 'Ten Week White' – very fragrant, good summer-flowering garden variety, 18 in (45 cm); 'Excelsior Mammoth Column White' – long-stemmed variety for cutting, large pots or garden border, 2½ ft (75 cm). For fully double flowers, select only seedlings with a notch in the seed leaves. Fertile, slightly alkaline soil in sun or partial shade. ✕

Matricaria eximia see *Tanacetum parthenium*.

Nicotiana affinis (tobacco plant)* – sticky leaves and chalk-white fragrant flowers which open in the evening; 3 ft (90 cm). 'Dwarf White Bedder' – compact bushy plant with scented flowers which remain open during the day; 16 in (40 cm). Rich well-drained soil in sun; does well in dry conditions.

*Nicotiana sylvestris** – stout plant with large light green fiddle-shaped leaves and clusters of semi-pendant, long-tubed fragrant flowers. Rich well-drained soil in a sheltered sunny position; 3–4 ft (90 cm–1.2 m).

Oenothera albicaulis (evening primrose) 'Mississippi Primrose' – fragile, slightly scented, cup-shaped flowers that open white and age to cream, then pink. Perennial but usually grown as a half-hardy annual. Ordinary soil in sun; 12 in (30 cm).

*Petunia hybrida grandiflora** – bushy plant with superb large, rather frilly flowers, not quite as floriferous as *multiflora*; 9–12 in (23–30 cm). Varieties include 'White Magic', F1 hybrid with pure white flowers 3½ in (9 cm) across and 'White Swan', F1 hybrid with frilly double flowers.

*Petunia hybrida multiflora** – smaller plant with numerous medium-sized trumpet-shaped flowers; 6–9 in (15–23 cm). Varieties include 'Prio White' which does well even in the poorest summer, and 'White Cloud', a good all-rounder.

Phlox drummondii – a small bushy phlox from Texas. Seed of the variety 'Dwarf Beauty' is available in separate colours, including white. Fertile soil in sun; 6 in (15 cm).

Portulaca grandiflorum (sun plant) – a succulent Brazilian species with reddish semi-trailing stems, fleshy bright green leaves and attractive yellow-stamened white flowers which resemble small semi-double roses; 'Swanlake' – large double white flowers. Excellent for bedding, edging and containers in dry sunny conditions; 4–6 in (10–15 cm).

Salvia farinacea 'White Porcelaine' – a fine border plant with neat sage-like leaves and dense white spikes. Ordinary soil in sun; 15 in (38 cm). ✕

Salvia splendens (bedding salvia) 'Carabinière White' – an ivory version of the dazzling scarlet bedder. Ordinary well-drained soil in sun; 14 in (35 cm).

Portulaca grandiflorum

Schizopetalon walkeri – a Chilean member of the cabbage family (Cruciferae) with ½–1 in (1–2.5 cm) fringed almond-scented flowers. Dislikes transplanting, so sow in peat pots. Good soil in sun; 6–12 in (15–30 cm).

Tagetes erecta (African marigold) 'Snowbird' – the best yet of varieties approaching white, with fully double, cream flowers. Any well-drained soil but best in a warm, sheltered sunny position; 18 in (45 cm).

Tanacetum parthenium (syn. *Chrysanthemum parthenium*, *Matricaria eximia*) (feverfew)* – a short-lived perennial with strong-smelling chrysanthemum-like foliage and small daisy flowers, usually treated as a half-hardy annual. 'Aureum' – bright yellow young foliage, single daisies, 18 in (45 cm) ✕; 'Golden Moss' (commonly known as golden feather) – a miniature with finely divided yellow foliage, 4 in (10 cm); 'Snow Dwarf' – double white pompoms with an outer ring of white guard petals, 9 in (23 cm); 'Tom Thumb White Stars' – similar in appearance and height to the previous variety; 'White Bonnet' – a tall border variety with double flowers, 2½ ft (75 cm) ✕; 'White Bouquet' – sprays of double mini-chrysanths, 2 ft (60 cm) ✕; 'White Gem' – ivory double pompom flowers surrounded by a circle of white ray petals, 8 in (20 cm). Dwarf varieties highly recommended for bedding, edging and containers. Ordinary well-drained soil in sun or partial shade. Tolerates dry shade.

Verbena x *hybrida* (bedding verbena) 'Marbella' – compact plant with neat pure white flowers. Good soil with added peat, in a sunny position; 8 in (20 cm).

Zinnia elegans – long-lasting, classy Mexican daisies; 'Big Snowman Double' – a dahlia-flowered type with huge rounded flowers, 2 ft (60 cm); 'Carved Ivory' – a cactus-flowered cultivar with equally large but more spiky cream blooms, 2½–3 ft (75–90 cm). Rich, well-drained soil in full sun. ✂

Tender Annuals

Browallia speciosa – pointed ovate leaves and 2 in (5 cm) five-petalled flowers. 'Silver Bells' – a good white winter-flowering variety; pinch out unless you want a semi-trailing plant, 18–24 in (45–60 cm); 'White Bell' – a bushy floriferous variety, reaching only 8 in (20 cm); 'White Troll' – a compact cultivar, 8–10 in (20–25 cm). May be sown in late summer for winter-flowering pot plants, or from winter to spring for summer flowers. Needs a minimum temperature of 55°F (13°C). Produces leaves at the expense of flowers if excessively fed and watered.

Exacum affine (Persian violet) – a naturally bushy compact species from the South Yemen island of Socotra, with neat shiny ovate leaves studded with beautifully fragrant 'violets'; 'Rosendal White'* – a free-flowering ball–shaped cultivar with yellow-eyed white flowers. Easy from seed sown either in late summer, giving bigger plants which must be over-wintered at 60°F (16°C), or in early spring; 6–9 in (15–23 cm).

Primula malacoides (fairy primrose) – a perennial Chinese primula, usually grown as an annual, with light green scalloped leaves and whorls of delicate primroses on slender stalks in winter and spring; 'Nordlicht' – a compact cultivar, reaching 8–10 in (20–25 cm), with single yellow-eyed white flowers; 'Snow Queen' – pure white single; 'Snow Storm' – pure white double. Peat-based compost at a minimum of 45°F (7°C); 12–18 in (30–45 cm).

Primula obconica – another Chinese primula which is perennial but usually grown as an annual; larger all round and bolder in effect than the former, with slightly hairy leaves (notorious for causing 'primula dermatitis') and heads of 1 in (2.5 cm) flowers; 'Giant White' – pure white; 'Juno White' – a compact

Exacum affine

F1 hybrid, reaching only 8 in (20 cm) and producing a succession of large flowers for much of the winter and spring. Cultivation as above; up to 15 in (38 cm).

Hardy Biennials

Cheiranthus cheiri (wallflower) – 'Ivory White' and 'White Dame', both have fragrant cream flowers in spring. Slightly alkaline, well-drained soil in sun or partial shade; 15 in (38 cm). ✂

Digitalis purpurea (foxglove) 'Alba'* – bold white spires, most effective in dappled shade. Expect a few purple-flowered plants among the whites; 3–5 ft (90 cm–1.5 m).

Geranium robertianum (herb Robert) – a succulent, unpleasant-smelling biennial that sows itself freely; rosettes of red-stalked deeply divided leaves that turn attractively red, and diminutive flowers from spring to autumn; 'Album' – brown-sepalled white flowers; 'Celtic White' – whiter flowers, bright green foliage and a dwarf compact habit; grows almost anywhere; 6–8 in (15–20 cm).

Hesperis matronalis (sweet rocket) 'Alba'* – fragrant four-petalled flowers in late spring (some plants bear pink-tinged flowers); 24 in (60 cm); 'Alba Plena' – double white (very rare now), needing vegetative propagation as it does not come true from seed; 'Candidissima' – a dwarf white form, 15 in (38 cm); best on

alkaline soil in sun or partial shade. Although biennial and dying after flowering, if the flower heads are removed before seed is set, the plant may develop sideshoots which can be used as cuttings.

Lunaria annua (honesty) 'Alba'* – white four-petalled flowers in spring followed by flat, silvery seed pods; 'Variegata'* – though capable of producing both purple- and white-flowered plants, some strains are almost wholly white-flowered and come remarkably true from seed. Mine has white-rimmed leaves and variegation, though scarcely apparent in seedlings, increases dramatically in the run up to flowering. Ordinary soil in partial shade; 2½ ft (75 cm).

Myosotis alpestris (forget-me-not) 'Alba'* – masses of tiny fragrant yellow-eyed flowers in spring; 'White Ball' – compact plants with slightly larger white flowers. Fertile, moisture-retentive soil in partial shade; 6 in (15 cm).

Oenothera acaulis (syn. *O. taraxifolia*) – perennial but treated as biennial. Prostrate stems of dandelion-like leaves and 3 in (7.5 cm) stalkless cup-shaped white flowers that open in the evening and turn pink as they fade. Ordinary soil in sun; 6 in (15 cm).

Oenothera acaulis

Oenothera taraxifolia see *Oenothera acaulis*.

Verbascum blattaria (moth mullein) 'Album' – spires of white flowers with fluffy red stamens. Ordinary soil in sun; 12–36 in (30–90 cm).

Verbascum hybridum (mullein) 'Mont Blanc'* – grey leaves and a branching spire of white flowers in summer. Ordinary soil in sun; can be propagated by root cuttings; 3–4 ft (90 cm– 1.2 m).

Hardy and Near Hardy Bulbs

Agapanthus africanus 'Albus' – strap-like leaves and dense rounded umbels of funnel-shaped flowers on tall bare stalks in summer. Forms handsome clumps when established. Well-drained soil in a shelterd sunny position, with protection in cold winters. This and the other species make excellent subjects for containers and tubs which can be brought under glass in hard winter weather; 18 in (45 cm). ✕

Agapanthus campanulatus 'Albus'* – a reasonably hardy species, similar in appearance to the former, but taller. Cultivation as above; 2–2½ ft (60–75 cm). ✕

Agapanthus orientalis see *Agapanthus praecox orientalis.*

Agapanthus praecox orientalis (syn. *A. orientalis*) 'Albiflorus' and 'Albus' – the white forms of a widely grown, variable species which is similar in appearance and cultivation requirements to the above; 2–2½ ft (60–75 cm). ✕

Allium cowanii – similar to *A. neapolitanum* but later and slightly taller.

Allium neapolitanum (syn. *A. cowanii*) (daffodil garlic) – 2 in (5 cm) heads of starry white flowers in spring. Not reliably hardy in cold areas. Well–drained soil in sun; 12–15 in (30–38 cm). ✕

Allium pulchellum 'Album' – a summer-flowering species with pointed flower buds and pyramidal heads. Cultivation as above; 12–18 in (30–45 cm). ✕

Allium rosenbachianum 'Album' – a stately plant with tightly packed starry cream flowers in spherical heads 5–6 in (12.5–15 cm) across. Cultivation as above; 2 ft (60 cm) or more. ✕

Allium stipitatum 'Album' – large round heads in late spring and early summer. Cultivation as above; up to 3 ft (90 cm).

Allium triquetrum – three-sided stalks and loose umbels of scented pendant white bells in early summer, each petal with a central green stripe. Not hardy in cold areas. Does well in partial shade; 15–18 in (38–45 cm).

Allium tuberosum (Chinese chives; garlic chives) – narrow, flat, garlic-flavoured leaves and white flowers in late summer. Well-drained soil in sun; 20 in (50 cm).

Allium ursinum (ramsons) – broad, rich-green leaves, rather like those of lily-of-the-valley in shape, smelling strongly of garlic, and starry umbels in late spring. Spreads well in bare damp shady places; 12 in (30 cm).

Allium ursinum

x *Amarygia parkeri* (syn. *Amaryllis belladonna* 'Parkeri') 'Alba' – stems of up to twelve trumpet-shaped flowers on bare stems in early autumn after the leaves have died down. Not hardy in cold areas. Plant at least 6 in (15 cm) deep in well-drained soil in a warm, sunny, sheltered position; 2 ft (60 cm).

Amaryllis belladonna 'Parkeri Alba' see X *Amarygia parkeri* 'Alba'.

Anemone appenina 'Alba' – white flowers with bluish undersides in early spring. Naturalises well in partially shaded grass or borders; 5 in (12.5 cm).

Anemone blanda 'White Spendour'* – large, long-lasting white flowers with lilac-tinged undersides in early spring. Good well-drained soil in sun or partial shade; 6 in (15 cm). ✕

Anemone coronaria 'De Caen The Bride' – milk-white petals and green centres. Well-drained soil in sun; 9–12 in (23–30 cm). ✕

Anemone nemorosa (windflower; wood anemone) 'Alba Plena' – the double white form; 'Hilda' – unusual flowers with a double circle of petals; 'Leeds Variety' – the largest white-flowered wood anemone, flushed pink on the reverse; 'Purity' – a floriferous white-flowered cultivar; 'Wilkes' White' – a good white variety. Moisture-retentive soil in light shade; 5 in (12.5 cm).

Anemone obtusiloba 'Alba' – a Himalayan anemone with neat white flowers in summer. Needs cool peaty soil and partial shade; 6 in (15 cm).

Arum italicum ssp. *albispathum* – handsome arrow-shaped leaves and inflorescences with cream spathes and white spadices. Ordinary soil in shade; 8–10 in (20–25 cm).

Camassia leichtlinii (quamash) 'Alba' – stout spikes of large creamy white stars; 'Semiplena' – the same but semi-double. Heavy moist soil with cool root run but heads in the sun; 2½–3 ft (75–90 cm).

*Cardiocrinum giganteum** – a giant Himalayan lily with heart-shaped leaves and huge white trumpets, maroon–streaked and smelling like honeysuckle and incense; 6–10 ft (1.8–3 m). Var. *yunnanense* – the Chinese form with blackish-purple stems, young bronze leaves, and ivory trumpets, green-tinted and opening from the top downwards; 7 ft (2.1 m). Both die after flowering but produce offsets which take 3–5 years to reach flowering size. Rich moist soil in woodland conditions.

Chionodoxa gigantea (glory of the snow) 'Alba' – several starry flowers per stem, large for the size of the plant, in early spring. Generally considered to be a large form of *C. luciliae*. Well-drained soil in sun; 6 in (15 cm).

Chionodoxa luciliae 'Alba' – another glory of the snow, with smaller white stars than the above and needing the same conditions; 4 in (10 cm).

Colchicum autumnale (meadow saffron; autumn crocus) 'Album'* – ungainly foliage during spring and summer, followed by generous numbers of large, fragile, crocus-like blooms

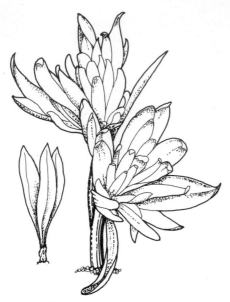

Colchicum autumnale *'Alboplenum'*

in early autumn. 'Alboplenum' is the double white form, which is not easy to come by. Colchicums do best in moisture-retentive, even heavy soil in sun or partial shade; 6 in (15 cm) – flowers; 12–16 in (30–40 cm) – leaves.

Colchicum speciosum 'Album'* – less pointed, more goblet-shaped flowers than the above species; size and cultivation the same.

Crinum x *powellii* (Cape lily) 'Alba' – a robust lily-like plant with untidy strap-shaped foliage and tall stems of large scented flowers in late summer; reasonably hardy in sheltered positions, though the new shoots may be badly damaged by frost unless protected; also good in tubs which can be taken under cover for the winter. Rich well-drained soil in sun; 2–3 ft (60–90 cm).

Crocus – all species and cultivars prefer well-drained soil and are best in sheltered position where the warmth of the sun will encourage the flowers to open even in very cold weather. Planting dwarf evergreen ground-cover, such as *Acaena* species, over the bulbs will prevent mud splashing in heavy rain.

Crocus albiflorus – white flowers, purple at the base, in early spring; cold damp spot; 3 in (7.5 cm).

Crocus biflorus (Scotch crocus) var. *parkinsonii* (syn. *C. biflorus* ssp. *biflorus*) – white inside with a dark yellow throat, striped purple and cream on the outside; var. *weldenii* 'Album' – pure white inside with grey-striped reverse; 4 in (10 cm).

Crocus chrysanthus 'Miss Vain' – pure white; 'Snow Bunting'* – white with purple feathering outside and an egg-yolk yellow centre; 3 in (7.5 cm).

Crocus fleischeri – early flowering species with starry white flowers and scarlet anthers; 3 in (7.5 cm).

Crocus hadriaticus – yellow-throated white flowers in autumn. Full sun; 3 in (7.5 cm).

Crocus 'Jeanne d'Arc'/'Joan of Arc'* – a large, pure white, Dutch hybrid which flowers in spring; 4–5 in (10–12.5 cm).

Crocus kotchyanus (syn. *C. zonatus*) 'Albus' – an autumn-flowering species from the Lebanon with pure white flowers, orange-spotted at the base. Propagated from a single plant found by George Barr in 1930; 3 in (7.5 cm).

Crocus ochroleucus – an autumn-flowering Syrian species with ivory to pure white flowers. Stands well in bad weather; 3–4 in (7.5–10 cm).

Crocus sativus var. *cartwrightianus* 'Albus' – the white form of a large variety of the saffron crocus; autumn flowering but not easy to grow without hot summers. Resents disturbance; 4 in (10 cm).

Crocus sieberi 'Bowles White'* – a robust early flowering white form of the Cretan crocus; 3 in (7.5 cm).

Crocus 'Snowstorm' – a large, pure white, Dutch hybrid with long-lasting goblet-shaped flowers in spring; 4–5 in (10–12.5 cm).

Crocus speciosus 'Albus'* – an easy autumn-flowering species; 5 in (12.5 cm).

Crocus tomasinianus 'Albus'* – vigorous, early flowering and excellent for naturalising; 3 in (7.5 cm).

Crocus vernus – parent of the ubiquitous large Dutch varieties; 'Albiflorus' – the white form is smaller than the species and flowers in late winter; 2–3 in (5–7.5 cm).

Crocus versicolor 'Picturatus' ('Cloth of Silver') – white with violet feathering (some plants may be pale mauve); 4 in (10 cm).

Crocus zonatus see *C. kotchyanus*.

Cyclamen coum 'Album' – plain green leaves (occasionally marbled) and white flowers with a pink 'nose' (pure white is very rare) in winter. Humus-rich soil in woodland conditions; 3 in (7.5 cm).

Cyclamen hederifolium (syn. *C. neapolitanum*) 'Album'* – silver-marbled leaves and masses of pure white flowers in late summer and early autumn. The white form is sometimes larger than the species. Cultivation as above; 4 in (10 cm).

Cyclamen neapolitanum see *Cyclamen hederifolium*.

Cyclamen repandum 'Album' – silver-marbled leaves with reddish undersides and scented white flowers with twisted petals in spring. Usually hardy if planted 3 in (7.5 cm) deep in a sheltered spot; 6 in (15 cm).

Endymion hispanicus see *Hyacinthoides hispanica*.

Endymion non-scriptus see *Hyacinthoides non-scripta*.

Erythronium dens-canis (dog's tooth violet) 'White Splendour'* – handsome brown-blotched foliage and white flowers in spring. Rich moist soil in partial shade; 6 in (15 cm).

Erythronium revolutum (trout lily) 'White Beauty' – white pagoda-shaped flowers above brown and white marbled leaves in late spring. Rich moist soil in partial shade; 12 in (30 cm).

Fritillaria meleagris (snake's head fritillary) 'Alba'* – green-chequered white bell-shaped flowers in late spring; 'Aphrodite'* – white. Damp soil in borders or grass; 10–12 in (25–30 cm). ✕

Erythronium revolutum

Galanthus (snowdrop) Avoid buying dormant bulbs which often fail to thrive and take years to become established. Snowdrops do best when transplanted in the spring when still green. They prefer damp shady sites.

Galanthus x *atkinsii* – a very tall and early hybrid with long pointed flowers; 10–12 in (25–30 cm). ✂

Galanthus 'Brenda Troyle'* – a large hybrid snowdrop with bold dark green markings and a noticeable fragrance; 6–10 in (15–25 cm). ✂

Galanthus elwesii – a tall species from Turkey with grey-green leaves and globular flowers. Needs sun; 6–10 in (15–25 cm). ✂

Galanthus nivalis (common snowdrop) – naturalises well in damp rich soil in partial shade; distinctive varieties include 'Flore Pleno' – increases enthusiastically and produces generous numbers of fat little double flowers; 'Viridipacis' – a vigorous plant with green-tipped outer petals in addition to the usual emerald crescents on the inner petals; 3–8 in (7.5–20 cm). ✂

Galanthus 'Samuel Arnott'* – a large hybrid snowdrop with slightly scented globular flowers on strong stems; 6–10 in (15–25 cm). ✂

Galtonia candicans (summer hyacinth) – tall spikes of slightly scented white bells in late summer. Plant 6 in (15 cm) deep in rich well-drained soil; 3–4 ft (90 cm–1.2 m). ✂

Hayacinthoides hispanica (syn. *Endymion hispanicus*; *Scilla hispanica*; *S. campanulata*) (Spanish bluebell) – 'Mount Everest', 'White City' and 'White Triumphator' are all tall, pure white varieties which flower in late spring and rapidly form large clumps. More vigorous and less graceful than the common bluebell. Partial shade; 12 in (30 cm).

Hyacinthoides non-scripta (syn. *Endymion non-scriptus*; *Scilla nutans*) (common bluebell) 'Alba' – graceful wild white bluebells in late spring. Partial shade; 12 in (30 cm).

Hyacinthus azureus see *Muscari azureum*.

Hyacinthus orientalis (hyacinth) – hardy but usually treated as half-hardy and kept frost-free for early flowering; often sold merely as 'White' but specialist bulb suppliers may offer 'L'Innocence'* or 'Carnegie', both fine whites, or 'Madam Sophie', a double white; very fragrant waxy flowers in winter (if 'prepared' bulbs are planted in early autumn) or spring. Pot-grown bulbs are best planted out in the garden after flowering and allowed to naturalise. Bedding hyacinths should be planted in the autumn, 5–6 in (12.5–15 cm) deep, in well-drained soil in sun or partial shade; 9–10 in (23–25 cm).

Hyacinthus orientalis 'Albulus' (Roman hyacinth) – more stems of fewer flowers than the florist's hyacinth, but sweetly scented and easy to grow; cultivation as above. ✂

Ipheion uniflorum 'Album' – a tough, easy member of the lily family, with garlic-smelling grassy leaves and 2 in (5 cm) starry white flowers in the spring. Well-drained soil in sun or partial shade; 7 in (17.5 cm).

Leucojum aestivum (summer snowflake) – tall stems with four to eight rounded snowdrop-like flowers in late spring; 'Gravetye Giant' is a superior form. Moist soil in partial shade; 2 ft (60 cm).

Leucojum autumnale – small white flowers, pink-flushed, in late summer; dry sunny position; 8–10 in (20–25 cm).

Leucojum vernum – a shorter, earlier flowering snowflake; 'Carpathicum' is a Rumanian form with yellow tips to the petals, instead of green. Damp soil in partial shade; 12 in (30 cm).

Leucojum vernum

Lilium auratum (golden-rayed lily of Japan)* – heavily scented open white flowers with a golden streak down the centre of each petal and raised crimson spots; 'Virginale' (also called 'Virginale Album') and 'Platyphyllum' have golden-banded petals but no spots. A stem-rooting, rather short-lived species which flowers in late summer and needs a sheltered spot in lime-free soil with shade at the base; spectacular in large pots; 4–6 ft (1.2–1.8 m). ✂

Lilium candidum (Madonna lily)* – scented, trumpet-shaped pure white flowers in early summer. Has a very short dormant period and must be planted by late summer when it produces an overwintering rosette of leaves; slightly alkaline soil in sun; 3–4 ft (90 cm–1.2 m). ✂

Lilium 'Casa Blanca' – an oriental hybrid with pure white *auratum*-type flowers in the summer; 3 ft (90 cm).

Lilium 'Green Magic' – a sophisticated white trumpet lily, shaded on the outside; 4–6 ft (1.2–1.8 m).

Lilium 'Imperial Silver' – an oriental hybrid with pure white fragrant flowers, finely dotted maroon; late summer; 5–6 ft (1.5–1.8 m).

Lilium martagon (Turk's cap lily) 'Albiflorum' – recurved white flowers with pink spots; 'Album'* – vigorous white form with twenty to thirty waxy flowers per stem; long-lived and tolerant but happiest on slightly alkaline soil in partial shade; naturalises well in grass or among shrubs; unpleasant scent; 2–4 ft (60 cm–1.2 m).

Lilium 'Mont Blanc' – an unscented Asiatic hybrid which bears wide-opening upright pure white flowers with a sprinkling of maroon dots in early summer. Easy in ordinary garden soil or pots of peat-based compost; 2 ft (60 cm). ✕

Lilium 'Olivia' – vigorous oriental hybrid with 6 in (15 cm) pure white flowers. Forces well; 2 ft (60 cm). ✕

Lilium regale (royal lily)* – a midsummer-flowering Chinese species with very fragrant trumpets, white inside with yellow throats, flushed wine-red outside; 'Album' – a robust form with pure white, yellow-throated flowers. An excellent reliable garden lily which is also superb in pots. Ordinary well-drained soil with its head in the sun and base sheltered by other plants; 4–6 ft (1.2–1.8 m). ✕

Lilium 'Simeon' – similar to 'Mont Blanc' but taller and with a pink flush in the centre; 2–3 ft (60–90 cm).

Lilium tigrinum (tiger lily) 'White Lady' – unscented recurved flowers in summer. Lime-free soil in sun; 3–4 ft (90 cm–1.2 m). ✕

Lilium wallichianum – a Himalayan species with long slender ivory trumpets; flowers late summer; 3–5 ft (90 cm–1.5 m).

Lilium 'White Happiness' – similar to 'Mont Blanc' but with ivory flowers. Good for forcing; 18–24 in (45–60 cm). ✕

Muscari azureum (syn. *Pseudomuscari azureum*; *Hyacinthus azureus*) 'Album' – tightly packed heads of tiny white flowers in early spring. Well-drained soil in sun; 5 in (12.5 cm).

Muscari botryoides (grape hyacinth) 'Album'* – paler, less vigorous foliage than the usual blue-flowered species and tight heads of pearl-like white bells in spring. Well-drained soil in sun 6–9 in (15–23 cm).

Narcissus – whether large or miniature, narcissi need good moisture-retentive soil in sun or partial shade. They should be planted in late summer and left to die down naturally (and untidily) after flowering. Do not knot or cut the leaves or you will get smaller and fewer flowers.

Narcissus 'Angel' – white with a small green-centred cup in late spring; 16–18 in (40–45 cm).

Narcissus 'Beersheba' – a vigorous white daffodil with a long trumpet and pointed petals; increases rapidly; 13 in (32.5 cm).

Narcissus cantabricus var. *foliosus* – milky white flowers in winter. Generally considered to be a variant of *N. bulbocodium*, the petticoat hoop narcissus. Best in a cold greenhouse on account of its early flowering and need for a warm dry summer rest; 5 in (12.5 cm).

Narcissus 'Cantatrice' – an elegant free-flowering white daffodil with relatively narrow petals and trumpet; 16 in (40 cm). ✕

Narcissus 'Dove Wings' – a spring-flowering *cyclamineus* hybrid with white petals and primrose cup fading to white; 14 in (35 cm). ✕

Narcissus 'Easter Moon'* – pure white short-cupped daffodil, green-tinted in the centre; resilient in bad weather, 18 in (45 cm). ✕

Narcissus 'Empress of Ireland' – an early flowering huge trumpet daffodil, 5 in (12.5 cm) across, with broad flat petals; 18 in (45 cm). ✕

Narcissus 'February Silver'* – long-lasting milky white miniature daffodils in early spring; 12 in (30 cm). ✕

Narcissus 'Ice Follies' – white with a flat frilly cream trumpet; increases well; 16–18 in (40–45 cm). ✕

Narcissus 'Jenny'* – a spring-flowering cyclamineus hybrid with pure white cup and white reflexing petals and narrow trumpet-shaped cup; 10–12 in (25–30 cm). ✕

Narcissus 'Moon Orbit' – a broad-petalled white daffodil which opens lime green and fades to ivory; 16–18 in (40–45 cm). ✕

Narcissus 'Mount Hood'* – probably still the

Narcissus *'Jenny'*

best white trumpet daffodil; opens cream and quickly turns white; vigorous and reliable; naturalises well; 16–18 in (40–45 cm). ✕

Narcissus 'Papillon Blanc' – an unusual 'orchid-flowered' white variety with a flattened split corona; 14 in (35 cm). ✕

Narcissus poeticus (pheasant's eye) 'Actaea'* – a large, tall cultivar with a bright red-margined yellow cup, excellent texture and heavenly scent; naturalises well and flowers very late (but earlier than other pheasant's eyes); 'Albus Plenus Odoratus' – pure white, loosely double, fragrant flowers; exquisite but shy-flowering until established; 'Cantabile' – green cup outlined in dark red; var. *recurvus* – very fragrant wild pheasant's eye with recurving petals; best naturalised in grass where it flowers long after other narcissi. All forms of *N. poeticus* do best planted deeply in damp heavy soil (including clay); 14–16 in (35–40 cm). ✕

Narcissus 'Rippling Waters' – a spring-flowering *triandrus* hybrid with two to three pure white flowers per stem; 14 in (35 cm). ✕

Narcissus 'Silver Chimes' – a neat late-flowering *tazetta* hybrid with dark leaves and up to ten scented white-petalled, cream-cupped flowers per stem. Naturalises well in dryish conditions; 12 in (30 cm). ✕

Narcissus 'Snowshill' – a perfectly simple large-cupped narcissus with a delightful scent; 16 in (40 cm). ✕

Narcissus 'Thalia' – a *triandrus* hybrid with cream petals and white cup; 12 in (30 cm). ✕

Narcissus 'Tresamble' – a tall fragrant *triandrus* hybrid with four to six open-cupped pure white flowers per stem in late spring; naturalises well; 14–16 in (35–40 cm). ✕

Narcissus triandrus (angel's tears) 'Albus'* – a spring-flowering Spanish form with clusters of one to six pendulous cream flowers which have globular trumpets and petals reflexed like a cyclamen; excellent for containers and the rockery; 6 in (15 cm).

Narcissus 'Vigil' – a large pure white trumpet daffodil with fine blue-green foliage; 18 in (45 cm). ✕

Narcissus 'White Cheerfulness'* – three or four fragrant double flowers, white shading to creamy yellow in the centre, on strong stems; naturalises and forces well; 16 in (40 cm). ✕

Narcissus 'White Marvel' – a sport from 'Tresamble' with double cups; 14 in (35 cm). ✕

Nothoscordum inodorum – an odourless relative of alliums with brown-striped white flowers in loose heads in spring. Well-drained soil in sun; 6 in (15 cm).

Ornithogalum arabicum – a handsome early summer-flowering star of Bethlehem whose white flowers have striking black pearl centres. Needs a warm, dry, sunny position; best in a cool greenhouse in cold areas; 18 in (45 cm).

Ornithogalum pyramidale – tall spikes of large white stars of Bethlehem in early summer. Leaves naturally begin to wither during flowering and dormancy follows soon after; 18–24 in (45–60 cm).

Ornithogalum nutans – spikes of green-streaked white flowers in late spring; needs a cool, shady position; becomes dormant quickly after flowering; 9 in (23 cm).

Ornithogalum umbellatum (star of Bethlehem) – many starry white flowers, green striped on the outside, in spring. Easy in ordinary well-drained soil but only opens in the sun; 6 in (15 cm).

Oxalis acetosella (wood sorrel) – pretty lime green clover leaves and fragile white flowers in spring. Nice on shady banks and in ferneries, enjoying peaty soil with added leafmould; 2 in (5 cm).

Oxalis deppei 'Alba' – bright green clover leaves and small funnel-shaped flowers in summer. Well-drained soil in sun or partial shade; 12 in (30 cm).

Oxalis enneaphylla – a species from the Falklands Islands with fragrant green-centred, white funnel-shaped flowers in summer, above crimped grey-green clover leaves; needs warmth and shelter; 3 in (7.5 cm).

Oxalis megallanica – a South American species with bronze clover leaves and white cups in late spring; 2 in (5 cm).

Oxalis magellanica

Pseudomuscari azureum see *Muscari azureum.*

Puschkinia scilloides (syn. *P. libanotica*) (striped squill) 'Alba' – similar to *Chionodoxa*, with five to eight flowers per stem in spring; slightly later than the blue. Nice in pots or in moist garden soil in sun or partial shade; 4–8 in (10–20 cm).

Puschkinia libanotica see *Puschkinia scilloides.*

Ranunculus ficaria (lesser celandine) 'Alba'* – shining ivory petals in early spring. Damp soil in partial shade; 3 in (7.5 cm).

Sanguinaria canadensis (bloodroot) – a rhizomatous woodlander from North America with short-lived, waxy, anemone-like flowers in spring as the exquisite lobed grey-green

leaves emerge; 'Flore Pleno'* – the larger longer-lasting double form. Worth every effort in well-drained but damp soil, enriched with peat and leafmould, preferably in the shelter of deciduous trees or shrubs. For best results plant 2–3 in (5–7.5 cm) deep, surrounded with sand; 6 in (15 cm).

Schizostylis coccinea (kaffir lily) 'Pallida'* – the palest form to date of this lovely autumn-flowering bulb, though it is surely only a question of time before a true albino is found. Spreads well in rich damp soil in a sheltered sunny position; 2–3 ft (60–90 cm). ✕

Scilla hispanica see *Hyacinthoides hispanica.*

Scilla nutans see *Hyacinthoides non-scripta.*

Scilla peruviana (Cuban lily) 'Alba' – neither Cuban nor Peruvian (though the first bulbs to reach England came on a ship named *Peru*), this Mediterranean species bears dense clusters of starry flowers in late spring or early summer. Plant shallowly in moist well-drained soil in a sheltered position in sun or partial shade; not reliably hardy in cold winters; 9–12 in (23–30 cm).

Scilla sibirica 'Alba' – a tougher spring-flowering species from Russia, with three to four stems per bulb of two to five bell-shaped flowers. Moist well-drained soil in sun or partial shade; 6 in (15 cm).

Trillium erectum (birthroot) 'Album' – a rhizomatous woodlander from North America with narrow greenish white three-petalled flowers above tri-lobed leaves in spring. Moist well-drained soil, enriched with peat and leafmould, in shade; 10 in (25 cm).

Trillium grandiflorum (wake robin)* – the showiest and easiest white trillium. Makes magnificent clumps when established; damp, leafmouldy soil in shade; 12 in (30 cm).

Trillium ovatum – similar to *T. grandiflorum* but smaller all round with flowers that age to pink. Conditions as above; 8–10 in (20–25 cm).

Tulipa (tulip) – all tulips should be planted 6 in (15 cm) deep in well-drained soil in a sunny position sheltered from strong winds. In general, they should be lifted as the foliage yellows and not allowed to decay in the soil where they may spread diseases which can ruin those planted the next year. The bulbs may be replanted in the autumn but few do as well the second year. 'Botanicals' such as *T. kaufmanniana* and its varieties, *T. biflora* and *T. turkestanica* make better perennials and can be left in the ground to ripen providing decaying foliage is removed.

Tulipa 'Angel' – unusual green and white Viridiflora tulip; 15–20 in (38–50 cm). ✄

Tulipa 'Athleet' – a Mendel hybrid with short rounded flowers mid-season; 15–20 in (38–50 cm). ✄

Tulipa biflora – clusters of small pointed flowers, ivory with yellow centres, in early spring; one of the few tulips which reliably flowers year after year though it may peter out gradually; 4 in (10 cm).

Tulipa 'Blizzard' – an early single pure white with oblong flowers and a sturdy habit; 14 in (35 cm). ✄

Tulipa fosteriana 'Purissima' (syn. 'White Emperor')* – brilliant long-lasting creamy-white blooms with yellow centres; 20 in (50 cm).

Tulipa kaufmanniana (waterlily tulip) 'Ancilla' – pinkish-red on the outside, opening to reveal primrose-shaded ivory inner petals with a rich golden base rimmed in red (almost too colourful to be counted as white!); 6 in (15 cm).

Tulipa 'Maureen' – a long-lasting oval-shaped ivory tulip which flowers in late spring; a favourite for bedding; 2 ft (60 cm). ✄

Tulipa 'Mount Tacoma'* – a strong late-flowering double with long-lasting peony-like white flowers; 16–20 in (40–50 cm). ✄

Tulipa polychroma – a very early species with several white, yellow-centred flowers to a stem. Needs a warm spot; 6 in (15 cm).

Tulipa 'Schoonoord' – an early double peony-flowered variety with a delicious honey scent; a sport of the old pink and white 'Murillo' and good for forcing if you can track it down; 12–15 in (30–38 cm).

Tulipa 'Snowpeak' – a large late-flowering ivory tulip with oval blooms; 27 in (67.5 cm). ✄

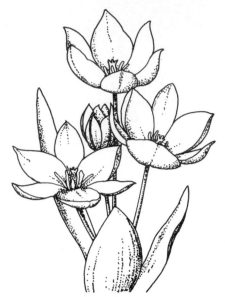

Tulipa polychroma

Tulipa 'Snow Queen' – an early double white shading to cream on the outside and yellow at the base; forces well; 11 in (27.5 cm).

Tulipa 'Snowstorm' – early, with short sturdy stems and fully double white flowers; 12 in (30 cm).

Tulipa 'Spring Green' – an unobtrusive but effective green and white Viridiflora tulip which flowers in late spring; 15 in (38 cm). ✄

Tulipa turkestanica – one of the very few tulip species with white flowers; five to nine small pointed flowers per stem in early spring; 8 in (20 cm).

Tulipa 'White Dream' – a good mid-season white with large weather-resistant blooms; 16 in (40 cm). ✄

Tulipa 'White Parrot'* – a late variety with ornately fringed petals; 18–24 in (45–60 cm). ✄

Tulipa 'White Triumphator'* – an unsurpassable late lily-flowered variety with grey-green leaves and curvaceous pure white flowers with pointed petals; 20 in (50 cm). ✄

Tulipa 'White Victory' – large late blooms which are jasmine-scented; 2 ft (60 cm). ✄

Zantedeschia aethiopica (arum lily; calla lily)* – sculptural white spathes in summer amid bright green arrow-shaped leaves. Excellent in large pots or deeply planted outdoors in damp rich soil in sun or partial shade, doing well in bogs, stream banks and shallow water; not reliably hardy in cold areas, where the hardier variety 'Crowborough' has more chance of success; 3 ft (90 cm) or more. ✄

Zephyranthes candida (flower of the west wind; Peruvian swamp lily) – fleshy grass-like leaves and slender stems of white crocus-like flowers in early autumn. Needs a warm sheltered position. Does well under glass in pots of John Innes No. 2 with additional sand, flowering better when pot-bound. 8 in (20 cm).

Half-hardy Bulbs
Acidanthera bicolor 'Murielae' see *Gladiolus callianthus*.

Anomatheca cruenta see *Anomatheca laxa*.

Anomatheca laxa (syn. *A. cruenta*; *Lapeyrousia laxa*; *L. cruenta*) – gladiolus-like leaves and graceful panicles of starry flowers in summer; 'Alba' – white; spreads by bulbils. Light well-drained soil in a sunny sheltered position (against a wall in cold areas); also suitable for pots in a cool conservatory; 7 in (17.5 cm).

Begonia evansiana 'Alba' – a tuberous oriental species with glossy red-backed leaves and delicate pendant 1 in (2.5 cm) white flowers in late summer; multiplies by bulbils in leaf axils; hardy in warm sheltered areas. Moist well-drained peaty soil in partial shade; 2 ft (60 cm).

Begonia x *tuberhybrida* (tuberous begonia) – white-flowered begonias are available in all the types used for containers and bedding, though they are seldom named anything but 'White': small double-flowered or 'Pendula' with a cascading habit; medium-sized double-flowered or 'Multiflora'; large double camellia-flowered, with blooms almost 6 in (15 cm) across; and double carnation-flowered or 'Fimbriata' with large fringed flowers. The tubers should be started into growth at room temperature in peat-based compost and set outside when all danger of frost is past; 12–24 in (30–60 cm).

Bletilla striata 'Alba' – an oriental orchid with a tuft of slender pleated leaves and graceful stems of about six pretty little orchid flowers, white with a pink-flushed lip and 2 in (5 cm) across, in late spring; hardy in warm sheltered areas; easy in pots for the cool conservatory if kept on the dry side during winter dormancy. Well-drained humus-rich soil in partial shade; 12 in (30 cm).

Dahlia – dahlia hybrids are easily grown and provide spectacular displays for beds and borders during the summer and autumn. The tubers should be planted 4 in (10 cm) deep in spring in rich well-drained soil in a sunny position. Removing dead flowers will prolong flowering. When frost has blackened the foliate, cut the plants down, leaving 6 in (15 cm) of stem, and carefully dig up the tubers. Place them upside down for a few weeks to dry off, then store at a minimum 40°F (5°C) until the following spring.

Dahlia 'Eveline' – a white 'decorative' with rounded petals and a pinkish-mauve centre; 3–3½ ft (90 cm–1.05 m). ✕

Dahlia 'My Love'* – a pure white 'semi-cactus' with spiky petals; 3–4½ ft (90 cm–1.35 m).

Dahlia 'Omo' – a single white 'lilliput', suitable for bedding and containers; 15 in (38 cm). ✕

Dahlia 'Polar Sight' – a giant 'cactus' with creamy-white shaggy petals; 4–5 ft (1.2–1.5 m). ✕

Dahlia 'Snowstorm' – a white 'decorative' with 6–8 in (15–20 cm) blooms; up to 4 ft (1.2 m). ✕

Dahlia 'White Aster' – a 'pompom' with immaculate 2 in (5 cm) golf balls; up to 4 ft (1.2 m). ✕

Eucomis bicolor (pineapple plant) 'Alba' – a rosette of broad, rather fleshy leaves and a pineapple-shaped spike of attractive but unpleasant-smelling starry flowers in late summer; may survive outdoors in mild areas if planted deeply and covered in winter, otherwise plant in spring in well-drained soil in sun and rescue in autumn; 12 in (30 cm).

Freesia x *kewensis** 'Ballerina' and 'Elder's Giant White' are two white-flowered cultivars of the hybrid florist's freesias which were made by crossing *F. refracta* with *F. armstrongii*; 'Diana' – the best double white, a favourite for bridal bouquets. Plant corms in rich sandy soil in a sunny sheltered position in the spring for summer flowering outdoors; for flowering

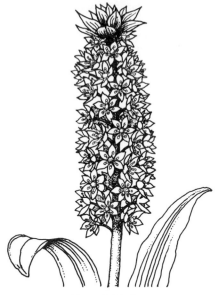

Eucomis bicolor

in the cool conservatory from late winter to spring, plant corms in pots of John Innes No. 2 in late summer and autumn, leaving outdoors until each has seven leaves; 12 in (30 cm). ✕

Freesia refracta 'Alba' – the species which is one of the parents of the multicoloured florist's freesias; small scented creamy-white flowers in summer, within six months of sowing seeds if kept in the greenhouse, otherwise from corms planted outside in rich sandy soil in full sun; 12 in (30 cm). ✕

*Gladiolus callianthus** (syn. *Acidanthera bicolor* 'Murielae') – an Ethiopian species with upside-down-looking purple-centred fragrant white flowers in late summer. Well-drained soil in full sun; may be planted outdoors in late spring and lifted in the autumn, or grown in pots for the greenhouse; 3 ft (90 cm). ✕

Gladiolus x *colvillei* 'The Bride' – 10 in (25 cm) spikes of smallish green-centred white flowers in summer. May be planted in autumn in sheltered areas, otherwise in spring; well-drained soil in sun; 18 in (45 cm). ✕

Gladiolus hybrids – stiff sword-shaped leaves and stately spikes of lily-like flowers with broad overlapping petals; 'Desire' – a small-flowered nanus hybrid with yellow-marked white flowers, 2 ft (60 cm); 'Ice Follies' – a butterfly hybrid with closely spaced medium-

sized ivory flowers which shade to cream in the throat, up to 3½ ft (1.05 m); 'White City' – a primulinus hybrid with smaller, more open spikes, 18–36 in (45–90 cm); 'White Friendship' – a vigorous giant-flowered white, 4 ft (1.2 m). Plant 4–6 in (10–15 cm) deep in well-drained rich soil in sun during the spring; lift corms in the autumn and keep dry and frost-free until the following spring. ✕

Gladiolus tristis – an African species with spikes of widely spaced scented yellowish-white flowers, sometimes red-tinged on the outside. Well-drained soil in sun; 18 in (45 cm).

Hymenocallis x *festalis* see *Ismene* x *festalis*.

Iris xiphioides hybrids (English iris) – similar to the above but with larger flowers in midsummer; 'Mont Blanc' – white with pink-spotted falls. Cultivation as above; 15–24 in (38–60 cm). ✕

Iris xiphium hybrids (Dutch iris and Spanish iris) – the well-known florist's iris can be grown from bulbs planted in spring. The Dutch flower in early summer, about a fortnight earlier than the Spanish, which need drier warmer conditions. Pure white cultivars, with a dash of yellow on the falls, are available, though rarely named anything but 'White'. Well-drained humus-rich soil in sun; 12–24 in (30–60 cm). ✕

Ismene x *festalis* (syn. *Hymenocallis* x *festalis*) (angel's lily; Peruvian daffodil) – strap-shaped leaves and a long-stalked umbel of large scented flowers with narrow curving petals and a daffodil-like trumpet. Plant 5–6 in (12.5–15 cm) deep outdoors in a well-drained sunny position in late spring to flower late summer, or in pots with the neck of the bulb above soil level during the winter to flower late spring; 18 in (45 cm).

Lapeyrousia cruenta see *Anomatheca laxa*.

Lapeyrousia laxa see *Anomatheca laxa*.

Lilium x *formolongii* 'White Swan' and 'Snow Trumpet' – cultivars of a cross between *L. formosanum* and *L. longiflorum*, the two finest half-hardy white trumpet lilies; seed sown in late winter produces plants which will flower the first summer; fragrant, waxy and long-lasting. Moist peaty compost in partial shade; 2–2½ ft (60–75 cm).

Lilium formosanum – a half-hardy lily from Taiwan with slender white scented trumpets, sometimes reddish on the outside, in late summer; var. *pricei* – a hardier variety, suffused brownish-maroon on the outside; produces its first flower within six months of sowing. Lime-free humus-rich soil in sun or partial shade; 4–6 ft (1.2–1.8 m).

Lilium longiflorum (Easter lily)* – an archetypal white lily with heavily scented trumpet-shaped flowers, 5–7 in (12.5–17.5 cm) long, in spring if planted in pots during the autumn and kept frost-free, or in late summer if planted 6 in (15 cm) deep outdoors in spring; 'Gelria' – three to eight flowers per stem, with rather fatter trumpets than the species; stem-rooting and lime-tolerant. Well-drained humus-rich soil in sun or partial shade; 3 ft (90 cm). ✂

Lilium speciosum 'Album'* – pure white fragrant recurved blooms in late summer; 'Kraetzeri' (syn. 'Roseum Album') – white with a green band down each petal and yellow rather than brown pollen; height and cultivation as above. ✂

Tigridia pavonia (tiger flower; peacock flower) – pleated gladiolus-like foliage and a succession of short-lived but magnificent triangular flowers, up to 5 in (12.5 cm) across, with vividly spotted saucer-shaped centres; 'Alba'* – white with dark red spots in the centre; 'Alba Immaculata' – unspotted white flowers, tending to greenish- or yellowish-white in some clones. Rich sandy soil in full sun, planting in spring and lifting in autumn; also good in large pots; 18 in (45 cm).

Tender Bulbs

Achimenes (hot water plant) – easy floriferous plants which grow from caterpillar-like tubercles. Plant three to five to a 6 in (15 cm) pot of peat-based compost in late winter at 60°F (16°C) and keep just damp until growth appears. Provide ample moisture, warmth and partial shade in the growing season, drying off in the autumn and storing in frost-free conditions when dormant. The tubers increase reliably from year to year. Flowering plants tend to drop open blooms if moved.

Achimenes dulcis – soft grey-green leaves and pure white trumpet-shaped flowers on trailing stems; 12 in (30 cm).

Achimenes x *hybrida* – slightly hairy ovate leaves and flat-faced tubular flowers on bushy semi-erect or trailing plants; 'Ambroise Verschaffelt' – a Victorian variety with purple-veined white flowers; 'Glacier' – a compact bushy cultivar with ice-white flowers; 'Jaureguia Maxima' (syn. 'Longiflora Alba') – an old variety with large white flowers on slender stems; 'Margaretta'/'Margarita' (syn. 'Longiflora Margaritacea') – an old trailing variety with light green foliage and pure white flowers; 'Moonstone'* – pure white flowers; 'White Giant' – an American cultivar with large white flowers marked with purple; 'White Knight' – large white flowers with tiny bright purple and yellow eyes; 'White Rajah' – a vigorous large white; 12 in (30 cm).

Albuca humilis – a South African species with linear leaves and vanilla-scented, yellow-centred, green and white striped flowers in ealy summer. Plant in pots of humus-rich sandy compost in early winter; needs a long dry dormant period after flowering; 6 in (15 cm).

Cyclamen persicum – a spring-flowering eastern Mediterranean species with silver-marbled leaves and very fragrant long-stalked white to pink flowers with dark pink 'noses'; 6–9 in (15–23 cm). Superseded in cultivation by the taller, larger-flowered (and often unscented) florist's hybrids and now available as tetraploids (such as 'By-Pass Pure White', 'Firmament Virgo' and the giant fringed 'Grandia') which will flower within eight months from seed. Mini-cyclamen, more like the species in terms of scent and flower size, have recently been introduced in a range of colours including pure white and pink-nosed white. Peat-based compost in moist cool conditions, around 55°F (13°C), with bright light but no direct sun.

Eucharis amazonica see *Eucharis grandiflora*.

Eucharis grandiflora (syn. *E. amazonica*) (Amazon lily)* – a South American species with long-stalked, large, dark evergreen ovate leaves and tall stems bearing clusters of pendant, highly scented pure white flowers, not unlike shallow-cupped daffodils in appearance; various flowering times, mostly in spring; enjoys ample warmth, humidity and water during the summer, with a drier lighter rest in winter at a minimum 55°F (13°C)

(though the foliage should never be allowed to wither). Peat-based compost with additional sand; 12–24 in (30–60 cm).

Habenaria radiata (rein orchid; butterfly orchid) – a terrestrial orchid whose fragrant delicate white flowers have bifurcated, deeply fringed lips. Sandy humus-rich compost in shade with plenty of warmth and water when in growth and minimum 55°F (13°C) during winter dormancy; 12 in (30 cm).

Haemanthus albiflos (syn. *H. carneus*) (blood lily) – an unusual summer-flowering South African species with two broad, wavy-edged, fleshy oblong leaves and a dense head of long-stamened white flowers. Plant during the autumn in pots of John Innes No. 2 with the neck of the bulb just covered and water moderately when in growth, keeping dry for several months after flowering; dislikes disturbance; needs a winter minimum of 50°F (10°C).

Haemanthus carneus see *Haemanthus albiflos*.

Hippeastrum hybrids (amaryllis) – stout fleshy stems with three to five huge flowers, each 7 in (17.5 cm) or more across, and strap-shaped leaves which appear during or soon after flowering. Named pure white cultivars include 'Christmas Gift', 'Dazzler', 'Mont Blanc', 'Nivalis' and 'Picotee' (the latter with a fine red line round each petal). Plant in peat-based compost with a third of the bulb exposed, in a pot only slightly wider than the bulb, and place in a warm sunny position. Keep on the dry side until growth begins, then water regularly. May be grown and flowered for years if fed and watered during the growing season, dried off for three months in the winter and stored at a minimum 55°F (13°C). Best results are obtained from bulbs measuring at least 12 in (30 cm) in circumference which should produce multiple flower stems; 12–18 in (30–45 cm).

Lycoris radiata 'Alba'* – a spectacularly beautiful flower, rather like a nerine but with much longer stamens. Plant with the neck of the bulb above the surface, in John Innes No. 2; may be grown outside in the summer and lifted in the autumn, but the delicate flowers are best appreciated in the shelter of a greenhouse; 16 in (40 cm). ✂

Narcissus tazetta (bunch-flowered narcissus)*

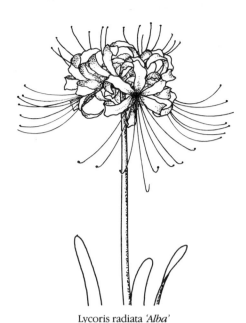

Lycoris radiata *'Alba'*

– clusters of fragrant flowers which open about six weeks after planting; 'Paper White' – the usual variety for forcing; var. *orientalis* – the Chinese New Year flower, often grown on pebbles in water. Easy and reliable as it needs no cold, dark period and may be placed straight on the windowsill after planting in pots of bulb fibre or bowls of pebbles and water; best treated as an annual; 12–15 in (30–38 cm). ✄

Nerine flexuosa 'Alba' – slender stems of up to a dozen crinkled lily-like flowers in autumn when the leaves have died down; needs a dry warm rest in summer to induce bud formation; may be hardy in very mild areas in well-drained soil in full sun but best in pots of John Innes No. 2 under glass; 18 in (45 cm). ✄

Ornithogalum thyrsoides (chincherinchee) – long-lasting white to cream flowers on tall spikes in summer; plant in pots of John Innes No. 2 in either autumn or spring for flowering in the cool conservatory in spring or summer respectively; 15 in (38 cm). ✄

Pamianthe peruviana – large evergreen bulbous plant, now very rare, with strap-shaped leaves 30 in (75 cm) long and a central cluster of five or six huge, very scented white flowers, rather like those of *Pancratium*. John Innes No. 3 with additional sand and a minimum winter temperature of 45°F (7°C). ✄

Pancratium maritimum (sea lily; sea daffodil) – a Mediterranean seashore bulb with grey-green leaves and umbels of very fragrant narrow-petalled flowers in the summer; may survive outdoors in mild areas if planted 8 in (20 cm) deep, but is better under glass and frost-free. John Innes No. 3 with additional sand, in full sun; 12 in (30 cm).

Pleione formosana 'Alba'* – a delightful orchid for shallow pots in the cool greenhouse; pure white blooms, about 3 in (7.5 cm) across and marked yellow inside the fringed tubular lip, emerge from green pseudobulbs in winter and early spring, followed by ribbed elliptic leaves. Plant in orchid compost so that the pseudobulbs sit on top of a mixture, keeping on the dry side until roots and leaves begin to grow; 6 in (15 cm).

Polianthes tuberosa (tuberose) 'The Pearl' – a Mexican jewel and one of the world's most fragrant flowers; long-lasting spikes of double flowers, each over 1 in (2.5 cm) across, are borne on tall stems in late summer. Treat as an annual in northern Europe and northern parts of North America as tubers virtually never build up to flowering size. Plant three or four to a 12 in (30 cm) pot in John Innes No. 1, keeping at 55–60°F (12–15°C) and on the dry side until leaves appear (they are slow); needs ample warmth, moisture and sunshine when in growth; 3–4 ft (90 cm–1.2 m). ✄

Sinningia speciosa (gloxinia) 'Mont Blanc' – rich green, oval, velvety leaves and huge pure white bells all summer; start in trays of damp peat at 70°F (20°C) in late winter, transferring to pots of peat-based compost when new growths are 1–2 in (2.5–5 cm); keep in a shady position at 64°F (18°C), gradually drying off as leaves die and keeping dry at 60°F (16°C) when dormant; 10 in (25 cm).

Hardy and Near Hardy Climbers

Actinidia chinensis (Chinese gooseberry; kiwi fruit) – a vigorous hardy deciduous twining climber with heart-shaped leaves, clusters of creamy-white flowers in summer, followed by the familiar hairy brown egg-shaped fruits. Male and female flowers are borne on separate plants so both are needed for fruit production. Needs a warm situation against a wall or sturdy fence/pergola; rich well-drained soil in sun or partial shade; up to 30 ft (9 m).

Actinidia kolomitka – a most elegant deciduous twining climber with pink- and white-tipped leaves which outshine the fragrant white flowers in early summer. Cultivation as above; 6–12 ft (1.8–3.6 m).

Bilderdykia baldschuanica (syn. *Polygonum baldschuanicum*) (Russian vine) – an alarmingly rampant deciduous twining climber which is covered in a froth of tiny ivory flowers (sometimes flushed pink) in late summer; grows over 10 ft (3 m) a year and is ideal for quickly covering eyesores. Any soil, any place; 40 ft (12 m).

Clematis – climbing species do best in alkaline soil and benefit from a mulch of well-rotted manure or compost each spring. They must have shade for the roots which can be provided by shrubs or paving stones. There is no rule of thumb for pruning as it depends on which group the plant in question belongs to: this should be determined when purchasing the plant.

Clematis akebioides 'Alba Luxurians' – green-tipped, mauve-tinted creamy white flowers. Tends to die back almost to ground level in winter and should be pruned to within 2 ft (60 cm) of the base; 8–12 ft (2.4–3.6 m).

Clematis alpina – a very hardy modest-sized species with a bushy habit and pendant flowers. There are a number of white varieties, including 'Burford White', 'Columbine White', 'Sibirica' (syn. 'Alba') and the semi-double 'White Moth'*; 6–8 ft (1.8–2.4 m).

Clematis alpina *'White Moth'*

Clematis apiifolia – a vigorous oriental species with trifoliate leaves and panicles of small white flowers in late summer and early autumn; cultivation as above; tolerates any aspect; 10 ft (3 m).

Clematis armandii – a vigorous evergreen species with large handsome glossy trifoliate leaves and long-stalked clusters of 2 in (5 cm) cream to blush-white flowers in spring; 'Snow Drift'* – larger pure white flowers. Cultivation as above; up to 30 ft (8 m) high and as much as 60 ft (18 m) across.

Clematis chrysocoma – a smallish deciduous species with distinctive brown-felted trifoliate leaves and small rounded pinkish-white flowers, rather like those of *C. montana*, from early summer to autumn; var. *sericea* (syn. *C. spooneri*) – a white-flowered, earlier, more vigorous variety which does well in shade; 10 ft (3 m).

Clematis 'Duchess of Edinburgh' – a summer-flowering double-flowered pure white hybrid; 10 ft (3 m).

Clematis fargesii souliei – a vigorous deciduous climber with compound leaves 9 in (23 cm) long and pure white flowers 2 in (5 cm) across throughout the summer; 20 ft (6 m).

*Clematis flammula** – a bushy deciduous species with bright green foliage and 12 in (30 cm) panicles of small scented white flowers from late summer to autumn; 10 ft (3 m).

*Clematis florida** – a rather sparse deciduous to semi-evergreen spring-flowering species which bears bold white flowers up to 3 in (7.5 cm) across, with a green ray down each petal; 'Plena' – double greenish-white flowers; 'Sieboldii' (syn. 'Bicolor') – white with striking purple staminodes; 10 ft (3 m).

Clematis 'Gillian Blades'* – an attractive summer-flowering hybrid with wavy-edged white flowers 7 in (17.5 cm) across; 10 ft (3 m) or more.

Clematis 'Henryi' – an old large-flowered hybrid with long pointed creamy-white sepals and dark stamens; blooms late spring to early summer, and again in late summer; 10 ft (3 m).

Clematis 'Huldine' – a late-flowering hybrid with pearly white flowers which have mauve bands on the reverse of the sepals; 10 ft (3 m).

Clematis 'Jackmannii Alba'* – the white form of one of the most popular hybrids ever, with a profusion of 4–5 in (10–12 cm) flowers in summer; deciduous; 10 ft (3 m).

Clematis 'John Huxtable'* – a late-flowering hybrid with pure white flowers; 8–10 ft (2.4–3 m).

Clematis macropetala – a slender bushy deciduous hybrid with pendant flowers, 2–3 in (5–7.5 cm) across, in early summer; 'Snowbird' – semi-double white; 'White Swan' – a lovely white form; 12 ft (3.6 m).

Clematis 'Marie Boisselot'* (syn. 'Madame le Coultre') – leathery leaves and broad-sepalled pure white flowers with cream bars and anthers; 10 ft (3 m).

Clematis montana – an ever-popular, easy and vigorous deciduous Himalayan species with cascades of four-sepalled white flowers, almost 2 in (5 cm) across, in late spring; 'Alba' – white; 'Alexander' – small creamy-white scented flowers; 'Grandiflora'* – large pure white; var. *wilsonii** – a late-flowering variety with large scented white flowers which have twisted sepals. Suitable for shady situations and ideal for growing into large trees and shrubs; about 25 ft (7.5 m).

Clematis rehderiana – a vigorous Chinese species with small cowslip-scented bell-shaped flowers, definitely cream rather than white, in late summer and autumn. Good for covering sheds and large eyesores; up to 25 ft (7.5 m).

Clematis spooneri see *C. chrysocoma* var. *sericea*.

Clematis vitalba (traveller's joy; old man's beard) – a rampant species with clusters of small four-sepalled creamy-white flowers in late summer, followed by silky seedheads. Suitable for the wild garden; up to 50 ft (15 m).

Clematis viticella 'Alba Luxurians' see *Clematis akebioides* 'Alba Luxurians'.

Clematis 'Wada's Primrose'* – a superb hybrid with yellow-anthered creamy white flowers in late spring/early summer; 7–10 ft (2.1–3 m).

Hydrangea anomala (syn. *H. altissima*) – a vigorous hardy deciduous root-climbing species with elliptic leaves and flat clusters of white flowers, some 7–8 in (17.5–20 cm) wide, in early summer; does best on a warm sunny wall. Well-drained moisture-retentive soil in a sheltered semi-shaded position that avoids early morning sun; 35 ft (10.5 m) or more.

Hydrangea altissima see *Hydrangea anomala*.

Hydrangea petiolaris – a vigorous hardy Japanese climber with pointed ovate leaves and flat clusters, up to 10 in (25 cm) across, of ivory flowers in early summer. Cultivation as above; 20–60 ft (6–18 m).

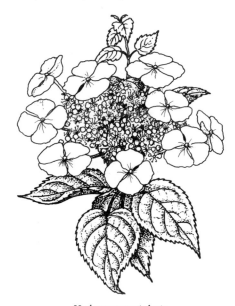

Hydrangea petiolaris

Jasminum officinale (jasmine)* – a vigorous hardy deciduous climber with slender stems of dark green pinnate leaves and sprays of heavenly-scented long-tubed white starry flowers in summer; 'Affine' – larger, pink-tinged flowers; 'Argenteovariegatum' – white-margined grey-green leaves; 'Aureovariegatum' – yellow-variegated foliage; 'Aureum' – golden leaves. Well-drained soil in sun; may need thinning out in spring but should not be cut back; 30 ft (9 m).

Lathyrus latifolius (everlasting pea) – an irrepressible deciduous climber which bears perfectly formed (but unscented) 1 in (2.5 cm) pea flowers on long stalks throughout the summer; ideal for interweaving through bush roses, hedges and shrubs to provide extra interest; 'Albus' – white; 'White Pearl'* – the best white form. Well-drained soil in sun; 6–10 ft (1.8–3 m).

Lonicera caprifolium (goat-leaf honeysuckle; perfoliate woodbine) – a deciduous climber distinguishable from *L. periclymenum* by its saucer-shaped perfoliate topmost leaves and rather earlier flowers; 'Cornish Cream'* – creamy-white scented flowers from early summer onwards. Well-drained humus-rich soil in sun or partial shade; 20 ft (6 m).

Lonicera japonicum (Japanese honeysuckle) – a vigorous evergreen or semi-evergreen twiner with thin trailing stems, dainty ovate leaves and small scented flowers in summer; 'Halliana' – white to cream flowers. Cultivation as above; 30 ft (9 m).

Lonicera periclymenum (honeysuckle; woodbine) – a deciduous twiner with ovate leaves and terminal heads of deliciously scented tubular flowers in summer; 'Graham Thomas'* – a floriferous cultivar with cream flowers which yellow as they age. Cultivation as above; 20 ft (6 m).

Passiflora caerulea (passion flower) – a reasonably hardy deciduous tendril climber with palmate foliage and complex flowers, sometimes followed by yellow fruits; 'Constance Elliot'* – a hardier form with pure white flowers. Well–drained humus-rich soil in a sunny sheltered position; 20 ft (6 m).

Polygonum baldschuanicum see *Bilderdykia baldschuanica*.

Schizophragma hydrangeoides – a deciduous summer-flowering climber from Japan, hydrangea-like, as the name suggests, and clinging by aerial roots; red-stalked hairy serrated heart-shaped leaves and 10 in (25 cm) flat heads of tiny flowers ringed by large ovate creamy-white bracts. Moisture-retentive humus-rich soil in sun or partial shade; 30 ft (9 m) or more.

*Schizophragma integrifolia** – a less vigorous Chinese version, stockier in appearance, with pointed ovate leaves and rather larger heads of flowers surrounded by white bracts over 3 in (7.5 cm) long. Cultivation as above; 20 ft (6 m).

Trachelospermum asiaticum – a rather tender evergreen self-clinging climber with dense leathery myrtle-like foliage and loose clusters of fragrant yellow-eyed ivory flowers (rather like those of periwinkles in shape) which age to cream. Light well-drained acid soil against a warm sunny wall; 15 ft (4.5 m).

Trachelospermum jasminoides – similar to the above species but with hairy new growths and white flowers; 'Variegatum' – foliage with ivory markings. Cultivation as above; 10–12 ft (3–3.6 m).

Trachelospermum majus – a vigorous species with foliage which tends to bronze in winter and pure white, faintly scented flowers. Cultivation as above; 20 ft (6 m).

Wisteria floribunda (Japanese wisteria) – the smaller of the two commonly grown species, distinguished by twelve to nineteen light green leaflets per leaf and slightly longer (on average) racemes of fragrant pea flowers in late spring; 'Alba'* – less vigorous than the species, with white flowers, occasionally lilac-tinted. Well-drained soil in a sunny sheltered position, away from structures which may be damaged by the vigorous root system and ever-thickening twining stems; 30 ft (9 m) or more.

Wisteria sinensis (Chinese wisteria) – a giant climber with up to eleven dark green leaflets per leaf and 8–12 in (20–30 cm) racemes of scented pea flowers in late spring; 'Alba'* – white flowers. Cultivation and warnings as above; may reach 100 ft (30 m).

Wisteria venusta – a Japanese species whose downy leaves have nine to thirteen leaflets ; short 4–6 in (10–15 cm) racemes of fragrant white flowers, each with a yellow spot at the base of the upright petal. Cultivation as above; 30 ft (9 m).

Tender Climbers

Beaumontia grandiflora (Nepal trumpet flower)* – a spectacular Indian twiner with shiny oblong-ovate leaves, rust-coloured stems and new growths, and large white datura-like flowers in summer. John Innes No. 2 with extra sand and peat; ample water, warmth and humidity during the summer and a minimum winter temperature of 60°F (15°C); 15–20 ft (4.5–6 m).

Bougainvillea glabra – the best species for pot culture, flowering well when young; twining stems, deciduous ovate leaves and panicles of insignificant flowers surrounded by flamboyant bracts; 'White Empress'* – pure white, occasionally flushed, or even reverting to magenta. Pot-grown plants need John Innes No. 3; cut back by about a third in early spring and maintain a winter minimum of 45°F (7°C); 6 ft (1.8 m) in a pot, 20 ft (6 m) or more in a border.

Calonyction aculeatum see *Ipomoea bonanox*.

Clianthus puniceus (parrot's bill) – an evergreen perennial sprawler from New Zealand with fine pinnate leaves and heavy clusters of dramatic creamy-white claw-shaped flowers in late spring and early summer; 'Albus' – the white form; 'White Heron'* – an especially good white form; John Innes No. 3 with additional sharp sand, well-ventilated conditions, and a winter minimum of 45°F (7°C); best trained onto a trellis; about 10 ft (3 m).

Cobaea scandens 'Alba'* (cathedral bells; cup-and-saucer vine) – a vigorous perennial tendril climber from Mexico, generally grown as an annual; pinnate leaves and numerous 3 in (7.5 cm) greenish-white bells, rather unpleasant smelling to attract bats, its pollinators in the wild. Easy from seed sown in spring at 65°F (18°C); needs a winter minimum of 40°F (5°C) to survive as a perennial; 20 ft (6 m) or more.

Hoya australis – a perennial evergreen climber with leathery ovate to roundish leaves and hanging umbels of pink-tinged white flowers scented like honeysuckle. Peat-based compost; warmth, humidity and partial shade during summer but on the dry side at a minimum temperature of 50°F (10°C) in the winter; 10 ft (3 m) or more.

Hoya carnosa (wax plant) – another Australian species, likewise perennial and evergreen; ovate leathery leaves, developing after new growth has reached its full length, aerial roots and dense umbels of fragrant waxy white flowers, pink-eyed and sometimes pink-tinged; 'Variegata' – cream-variegated, pink-flushed foliage. Cultivation and height as above.

Hoya multiflora – a twiner from Borneo which bears almost veinless leathery oval leaves and drooping umbels of numerous waxy white-nosed shooting star flowers (again, not truly white as the backward-pointing segments are yellowish). Cultivation as above but with rather higher temperatures and humidity.

Ipomoea alba see *Ipomoea bona-nox*.

Ipomoea bona-nox (syn. *I. alba; Calonyction aculeatum*) (moon flower)* – a tropical American perennial bindweed whose flowers open at dusk and close at dawn; light green heart-shaped leaves and a succession of giant soap-scented pure white 'morning glory' flowers all summer under glass. Quick and easy from seed sown in late winter at 75°F (25°C); best treated as an annual; 10 ft (3 m) or more.

Jasminum polyanthum (jasmine)* – a semi-evergreen half-hardy Chinese species commonly grown as a pot plant; dark green pinnate leaves and cascades of very fragrant white or blush-white flowers from winter to spring, according to temperature. Peat-based compost, sun and a winter minimum of 40°F (5°C); avoid excessive water and humidity in summer so that the growths can ripen for flowering; up to 10 ft (3 m).

Jasminum sambac (Arabian jasmine; sambac; maid of Orleans)* – a perennial evergreen Indian species, sprawling rather than climbing in habit, with large intensely fragrant flowers which may be produced most of the year; 'Grand Duke' – double. Needs ample warmth in summer and a winter minimum of 60°F (15°C); 5 ft (1.5 m).

Lapageria rosea (Chilean bellflower) – a slender twining climber with leathery pointed ovate leaves and pendant waxy elongated bells, 3 in (7.5 cm) long, in summer; 'Albiflora'* – pure white flowers; does best in lime-free compost with additional peat and leafmould; keep evenly moist and at a winter minimum of 45°F (7°C); up to 15 ft (4.5 m).

Mandevilla laxa see *Mandevilla suaveolens*.

Mandevilla suaveolens (syn. *M. laxa*) (Chilean jasmine) – a deciduous tropical American twiner with dark green heart-shaped leaves and fragrant pure white petunia-like blooms in spring and summer. Unfortunately it is not very successful in pots but does well in a greenhouse border; peat-based compost with additional sand; warmth, humidity and sun in summer, with a minimum winter temperature of 55°F (13°C); 15 ft (4.5 m) or more.

Pandorea jasminoides (bower plant) – an evergreen Australian twiner with glossy compound leaves and 2 in (5 cm) pink-throated white flowers in summer and autumn; variable, with some forms pure white and others decidedly pink. John Innes No. 3, full sun and good ventilation with a winter minimum temperature of 45°F (7°C); 10–20 ft (3–6 m).

Passiflora edulis (passionflower)* – an evergreen tropical American climber; three-lobed leaves and fascinating fringe-filled 3 in (7.5 cm) white flowers which have plum-coloured centres and are followed by luscious passion fruits. May succeed in a large pot or tub in John Innes No. 3; needs plenty of warmth, moisture and humidity in summer when flowering and fruiting, followed by drier conditions at a minimum temperature of 50°F (10°C) in winter; up to 20 ft (6 m).

Plumbago capensis 'Alba' – the white-flowered form of a vigorous evergreen South African climber; light green elliptic leaves and panicles of long-tubed flowers from spring to autumn, resembling jasmine in appearance but scentless. May be grown in pots of John Innes compost No. 3 but is best in a greenhouse border; feed and water generously during the summer but keep on the dry side and at a minimum of 45°F (7°C) in winter; 15 ft (4.5 m).

Solanum jasminoides (jasmine nightshade) 'Album'* – the lovely white-flowered form of a twining evergreen climber from Brazil; shiny pinnately lobed light green foliage and branched clusters of yellow-anthered star-shaped white flowers all summer. John Innes No. 2, plenty of water and good ventilation in summer; on the dry side at a minimum temperature of 40°F (5°C) during the winter; 10 ft (3 m) or more.

Stephanotis floribunda (bridal bouquet)* – an evergreen Madagascan twiner with leathery oblong leaves and clusters of deeply fragrant tubular waxy white flowers in summer. Peat-based compost with shade, ample warmth, moisture and humidity during the summer, and a dryish sunny rest in winter at a minimum temperature of 55°F (13°C); 12 ft (3.6 m). ✕

Thunbergia erecta (bush clock vine) 'Alba' – an evergreen West African twiner with smooth dark green ovate leaves and yellow-tubed white flowers for most of the year. John Innes No. 2 with ample warmth in the growing season and a winter minimum temperature of 55°F (13°C); 6 ft (3 m).

Thunbergia fragrans – an evergreen tropical Asian twiner with toothed, slightly downy ovate leaves and pure white, lightly scented flowers marked sparingly with yellow in the throat; flowers in about four months from seed and can therefore be treated as an annual. Cultivation as above; 8–10 ft (2.4–3 m).

Thunbergia grandiflora (sky vine) 'Alba' – the white-flowered form of a quick-growing vigorous evergreen twiner from northern India; ovate, slightly downy leaves and abundant 2–3 in (5–7.5 cm) flared tubular flowers all summer. Cultivation as above; 20 ft (6 m) or more.

Trachelospermum (see list of hardy and near hardy climbers, page 113).

Hardy and Near Hardy Perennials

Achillea ageratifolia – dense mounds of silvery leaves and white flowers in midsummer. Well-drained soil in sun; 6 in (15 cm).

Achillea ageratum (sweet Nancy; sweet Maudlin) – close, finely cut silver leaves and yarrow-like flowers in summer. Cultivation as above; 6–8 in (15–20 cm).

Achillea argentea see *A. clavennae*.

Achillea clavennae (syn. *A. argentea*) – deeply cut silver leaves and loose heads of white flowers in summer; 'Integrifolia' – untoothed leaves. Cultivation as above; 6–8 in (15–20 cm).

Achillea decolorans 'W. B. Child' – loose flat heads of white flowers in early summer with another flush in autumn. Cultivation as above; 2 ft (60 cm). ✕

Achillea ptarmica (sneezewort) – small narrow-toothed leaves and loose clusters of white daisy-like flowers throughout the summer; 'The Pearl'* – double white buttons; 'Perry's White' – similar but earlier. Rather invasive, tolerates quite wet conditions. Ordinary soil in sun; 2½ ft (75 cm). ✕

Achillea x *wilczekii* – broad silver leaves and white flowers in summer. Ordinary soil in a dry sunny position; 8 in (20 cm).

Aconitum x *cammarum* 'Grandiflorum Album' – a large white-flowered monkshood. Moist humus-rich soil in partial shade; 3½ ft (1.05 m).

Aconitum napellus (monkshood) – a poisonous species with deeply cut leaves and racemes of hooded flowers in summer; 'Albidum' – white. Moist humus-rich soil in partial shade; 3½ ft (1.05 m).

Aconitum septentrionalis 'Ivorine'* – deeply cut foliage and spikes of small hooded ivory flowers in late spring. All parts poisonous. Moist humus-rich soil in semi-shade; 2–3 ft (60–90 cm). ✕

Actaea alba see *A. pachypoda*.

Actaea pachypoda (syn. *A. alba*) (baneberry; herb Christopher; toadroot)* – clumps of elegant toothed pinnate leaves, fluffy white flowers and poisonous crimson-stalked white berries, each smartly finished with a black dot. Moist humus-rich soil in partial shade; 12–18 in (30–45 cm).

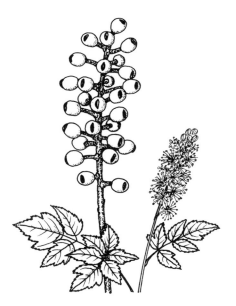

Actaea pachypoda

Ajuga reptans (bugle) 'Alba'* – shiny bronze–flushed rosettes of leaves and stocky spires of white flowers in late spring; excellent ground-cover for damp shady conditions; 4 in (10 cm).

Anaphalis cinnamomea (syn. *A. yedoensis*) – cinnamon-scented dark grey silver-edged leaves and pearly white everlasting daisies in early autumn. Well-drained to dry soil in sun; 2½ ft (75 cm). ✕

Anaphalis margaritacea – grey leaves and numerous bunches of white everlasting flowers in late summer. Cultivation as above; 12–18 in (30–45 cm). ✕

Anaphalis nubigena – a bushy plant with narrow woolly grey leaves and clusters of off-white everlasting flowers in late summer. Cultivation as above; 6–8 in (15–20 cm). ✕

Anaphalis triplinervis (pearl everlasting) – silver downy leaves and bunched clusters of papery white flowers in late summer; 12 in (30 cm); 'Summer Snow' – a compact form, 10 in (25 cm). Cultivation as above. ✕

Anaphalis yedoensis see *A. cinnamomea*.

Anemone x *hybrida* (syn. *A. japonica*) (Japanese anemone) – long-stalked autumn-flowering anemones; 'Alba' – white; 'Géante des Blanches' (syn. 'White Giant'; 'White Queen')* – a vigorous large-flowered white; 'Honorine Jobert' – tall white; 'Louise Uhink' – a large single white. Moisture–retentive soil in partial shade; 3–4 ft (90 cm–1.2 m).

Anemone japonica see *Anemone* x *hybrida*.

*Anthemis cupaniana** – a quick-growing sprawling plant with finely cut silver foliage and large classic white daisies in early summer; excellent ground-cover. Well-drained soil in sun; 15 in (38 cm). ✕

Anthemis nobilis see *Chamaemelum nobile*.

Anthericum liliago major (St Bernard's lily) – tufts of grey-green grassy leaves and spires of starry white flowers in late spring and early summer. Light moisture-retentive soil in partial shade; 18–24 in (45–60 cm). ✕

Aponogeton distachyos (water hawthorn)* – an aquatic with elongated oval floating leaves and forked spikes of black-anthered waxy white flowers which are long-lasting and heavily scented; flowers from early summer to autumn; tolerates shade but flowers better in sun. Neutral to slightly acid soil, still or slow-moving water 6–18 in (15–45 cm) deep. ✕

Aquilegia flabellata – a miniature Japanese columbine with grey-green dissected leaves and proportionately large short-spurred flowers in early summer; 'Alba' – white; 'Nana Alba' – a compact white form. Moist humus-rich soil in sun or partial shade; 6 in (15 cm).

Aquilegia fragrans – a Himalayan species with large scented white flowers in summer. Cultivation as above but in a very sheltered position; 12 in (30 cm). ✕

Aquilegia vulgaris (columbine; granny's bonnet) 'Alba' – single white; 'Clematiflora Alba' (syn. 'White Spurless') – double white; var. *nivea* (syn. 'Munstead White')* – grey-green foliage and short-spurred flowers; 'Snow Queen' – a pure white long-spurred columbine. Cultivation as above; 2½ ft (75 cm). ✕

Arabis albida see *A. caucasica*.

Arabis caucasica (syn. *A. albida*) 'Flore Pleno' – quick off the mark but less invasive than the species, with grey-green leaves and double flowers in spring; 'Variegata' – brilliant white single flowers and cream-edged leaves. Well-drained soil in partial shade; 5 in (12.5 cm).

Arabis ferdinandii–coburgii 'Old Gold' – compact tufts of golden variegated leaves and white flowers in spring; 'Variegata' – white-variegated foliage; cultivation as above; 4–6 in (10–15 cm).

*Arenaria montana** – dense mats of narrow dark green leaves and masses of little white flowers in spring; superb for rockeries, walls and path edges. Well-drained soil in sun or partial shade; 3 in (8 cm).

Armeria maritima (thrift) 'Alba' – neat evergreen hummocks of grassy leaves and drumstick flowers in spring. Ordinary soil in sun; 6 in (15 cm).

Artemisia lactiflora (white mugwort) – an Asian species with deeply cut, dark green leaves and plumes of tiny scented off-white flowers in late summer. Moisture-retentive soil in partial shade; 2–4 ft (60 cm–1.2 m).

Aruncus aethusifolius – finely divided leaves and dense plumes of minute cream flowers in early summer, followed by good autumn colour. Rich moist soil in partial shade; 9–12 in (23–30 cm).

Aruncus dioicus (syn. *A. sylvester*) (goat's beard) – a handsome plant for rich moist soil, rather like meadowsweet but larger and with more elegant cream plumes in early summer; 'Kneiffii' is the same but only half the size; 4–6 ft (1.2–1.8 m).

Aruncus plumosus 'Glasnevin' – similar to the above but smaller and neater; 4 ft (1.2 m).

Aruncus sylvester see *A. dioicus.*

Asperula odorata see *Galium odoratum.*

Asphodelus albus (white asphodel) – clumps of narrow-keeled leaves and stout stems with terminal spikes of starry flowers in summer, each petal marked with a soft brown stripe. Poor dry soil in full sun; 2–4 ft (60 cm–1.2 m).

Asphodelus albus

Aster alpinus 'Albus' – a dwarf spreading aster with grey-green leaves and large yellow-centred white daisies in summer. Well-drained soil in sun; 6 in (15 cm).

Aster corymbosus see *A. divaricatus.*

Aster divaricatus (syn. *A. corymbosus*) – contrasting shiny black stems and sprays of diminutive pink-tinted daisies in late summer. Cultivation as above; 2 ft (60 cm). ✂

Aster dumosus 'Kristina' – compact dark green foliage and double daisies. Cultivation as above; 15 in (38 cm). ✂

Aster ericoides 'White Heather' – slender branched stems of tiny white daisies in autumn. Cultivation as above; 2–3 ft (60–90 cm). ✂

Aster novae-angliae (Michaelmas daisy) 'Autumn Snow' (syn. 'Herbstschnee')* – tall bushy stems of superb white daisies in autumn. Cultivation as above; 4 ft (1.2 m) or more. ✂

Aster novii-belgii (Michaelmas daisy) – 'Snowsprite' – the best dwarf white, 9 in (23 cm); 'White Ladies'* – a robust white; 'White Swan' – excellent early white. Cultivation as above; 4 ft (1.2 m). ✂

Astilbe x *arendsii* – divided leaves and erect plumes of tiny flowers in summer; 'Irrlicht' – dark leaves and creamy-white flowers; 'Snowdrift'* – bright green divided leaves and pure white plumes; 'White Gloria' – creamy-white plumes. Rich moist soil in sun or shade; 2 ft (60 cm).

Astrantia major (masterwort) – lobed leaves with serrated margins and long-lasting papery flowers in summer; 'Alba' – larger, whiter petals than the species, but with the same attractive greens tips and undersides. Ordinary soil in partial shade; 24–30 in (60–75 cm)

Balsamita major (syn. *Balsamita vulgaris*; *Chrysanthemum balsamita*) var. *tomentosum* (camphor plant) – silky grey-green lanceolate leaves with a camphoraceous smell, and clusters of small white daisies in summer; spreads quickly into large clumps. Ordinary soil in sun; 3 ft (90 cm) or more.

Balsamita vulgaris see *Balsamita major.*

Bellis perennis (common daisy) 'White Carpet' – neat little plants with perfect double white daisies. Moisture-retentive soil in sun or partial shade; 4–6 in (10–15 cm).

Bergenia 'Bressingham White'* – a free-flowering vigorous white. Ordinary soil in sun or partial shade; 12 in (30 cm).

Bergenia stracheyi 'Alba' – a small Himalayan bergenia with neat oval leaves and clusters of white bells which have green calyces; 'Silberlicht' – slightly larger and with pink calyces. Cultivation as above; 6–8 in (15–20 cm) and 12–15 in (30–38 cm) respectively.

Boltonia asteroides (false chamomile) – a tall North American daisy; 'Snowbank' – white daisies in late summer. Ordinary moisture-retentive soil in sun or partial shade; 4–5 ft (1.2–1.5 m). ✂

Caltha palustris (marsh marigold) 'Alba'* – the white-flowered Himalayan form with fewer, smaller flowers and a liking for damp, rather than aquatic, conditions; 6 in (15 cm).

*Campanula alliariifolia** – grey-green heart-shaped leaves and tapered spires of narrow cream bells in summer; perfect with old roses; 'Ivory Bells' – larger flowers. Well-drained soil in sun or partial shade; 18–24 in (45–60 cm).

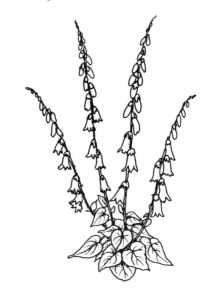

Campanula alliariifolia

Campanula barbata 'Alba' – a delightful dwarf species whose large bells have intriguing whispy beards. Cultivation as above; 3–12 in (7.5–30 cm).

Campanula carpatica 'Alba' and 'Bressingham White'* – wonderful free-flowering plants, smothered with large upturned white bells in summer; 'Snowsprite' – late-flowering and dwarf; 'Turbinata Alba' – grey hairy foliage, a compact habit and bluish-white bells; excellent for the edges of paths and borders or over rocks and walls. Cultivation as above; 9–12 in (23–30 cm).

Campanula cochlearifolia 'Alba'* – the dearest little miniature white harebell; easy and free-flowering in containers or somewhere it will not be bullied. Cultivation as above; 3–4 in (7.5–10 cm).

Campanula glomerata – dense clumps of foliage and stiff stems of clustered bells in summer; 'Alba' – white; 'Crown of Snow' – pure white. Cultivation as above; 15 in (38 cm).

Campanula lactiflora – tall branched stems of open bells in summer; 'Alba'* – white; 'White Pouffe' – a dwarf cultivar which makes green hummocks studded with open white bells. Cultivation as above but prefers full sun; up to 5 ft (1.5 m).

Campanula latifolia (giant bellflower) – 'Alba' and 'White Ladies'* – bold spikes of tightly clustered large, narrow bells. Moisture-retentive soil in partial shade; 3–4 ft (90 cm–1.2 m).

Campanula latiloba 'Alba'* – a Siberian species with narrow serrated leaves and broad, outward-facing bells in early summer. Well-drained soil in sun or partial shade; 18 in (45 cm).

Campanula persicifolia (peach-leaved bellflower) – rosettes of long, narrow, leathery leaves and spikes of cup-shaped bells; one of the loveliest bellflowers for the border; 'Alba'* – white; 'Boule de Neige' – double white; 'Planiflora Alba' – a dwarf white-flowered form. Cultivation as above; 12–20 in (30–50 cm).

Cardamine asarifolia see *Pachyphragma macrophylla.*

Catananche caerulea (cupid's dart) 'Perry's White' – narrow grey-green leaves and long-lasting cornflower-like blooms on wiry stems. Sunny, well-drained position; 18–30 in (45–75 cm). ✂

Centaurea montana 'Alba' – a tough perennial white cornflower which flowers in early summer. Well-drained soil in sun; 18–24 in (45–60 cm).

Centranthus ruber (red valerian) 'Albus'* – grey-green oval leaves and rounded panicles of tiny flowers throughout the summer; grows happily in walls or borders. Poor, well-drained soil, preferably alkaline; 18–36 in (45–90 cm).

Cerastium tomentosum (snow-in-summer) – a rampant carpeter with grey leaves and chalk-white flowers; 'Columnae' – a dwarf restrained form of the space invader, with silvery foliage and pretty little white flowers in early summer. Well-drained soil in sun; 3 in (7.5 cm).

Chamaemelum nobile (syn. *Anthemis nobilis*) (Roman chamomile; lawn chamomile) – feathery aromatic foliage and small, slightly droopy yellow-centred daisies in summer; 'Flore Pleno'* – dishevelled creamy-white double flowers; lovely for the sides of paths and around seats. Well-drained soil in sun; 6–9 in (15–23 cm).

Chrysanthemum balsamita see *Balsamita major.*

Chrysanthemum coccineum (syn. *Pyrethrum roseum*) – bright green feathery leaves and large yellow-centred daisies in summer; 'Avalanche' – single white; 'Carl Vogt' and 'Mont Blanc' – double whites; 'Silver Challenger' – double white. Ordinary soil in sun; 3 ft (90 cm). ✂

Chrysanthemum x *hybridum* (probably to be transferred to *Dendranthemum*) (florists' chrysanthemum) – all varieties of chrysanthemum, from small single daisies to those grown for exhibition, are available in white. They include: 'Arctic Snow' – pure white single, 4 ft (1.2 m); 'Blanche Poitevene' – pure white reflexed, 3 ft (90 cm); 'Fred Shoesmith' – cream-centred white decorative, popular for Christmas cut flowers, 4 ft (1.2 m); 'Himalaya' – huge ball-shaped pure white incurved blooms, 3½ ft (1.05 m); 'Maylen' – 6–7 in (15–17.5 cm) incurved ivory blooms, 4 ft (1.2 m); 'Pavlova' – curly reflexed blooms, 3 ft (90 cm); 'Pennine Dove' – a single spray chrysanth; 'Pennine Ski' – double pure white spray; 'Polar Gem' – green-tinged white incurved blooms, 4½ ft (1.35 m); 'Seagull' – a cascade (charm) chrysanth with numerous single white daisies, like a giant bouquet, 2 ft (60 cm) high and 3 ft (90 cm) across; 'Snow Princess' – a small-flowered white pompom variety which will flower well in the open from late summer until the first frosts, 18 in (45 cm); 'White Harry Gee' – a

giant-flowered Japanese exhibition cultivar with pure white coiffeured blooms, 4 ft (1.2 m). Rich, moisture-retentive, well-drained soil in sun. Though hardy, late-flowering and exhibition varieties need protection from the elements. ✂

Chrysanthemum maximum see *Leucanthemum maximum.*

Chrysanthemum nipponicum – clumps of large white daisies in autumn; needs a warm position; 15 in (38 cm). ✂

Chrysanthemum uliginosum (syn. *Leucanthemella serotina*) (Hungarian daisy; moon daisy) – tall sprays of green-centred white daisies in autumn; 7 ft (2.3 m). ✂

Cimifuga cordifolia – slender branched spikes of tiny fluffy white flowers in late summer. Moist humus-rich soil in partial shade; 4 ft (1.2 m).

Cimifuga foetida see *C. simplex.*

Cimifuga racemosa (black snake root) – delicate feathery white spikes in summer. Cultivation as above; 5 ft (1.5 m).

Cimifuga ramosa – a tall Japanese species with narrow bottle-brush spikes of tiny white flowers in late summer; 'Atropurpurea' – purplish-black stems and dark foliage. Moist humus-rich soil in light shade; 5–6 ft (1.5–1.8 m).

Cimifuga simplex (syn. *C. foetida*) (bugbane) 'Elstead' – purplish-green stems and slender, curved bottle-brushes which are white with pink stamens; 'White Pearl' – pure white; autumn-flowering. Cultivation as above; 4 ft (1.2 m).

Clematis recta – a bushy species suitable for the border; small fragrant white flowers in summer, followed by silky seedheads; needs support. Ordinary well-drained soil in sun; 3–4 ft (90 cm–1.2 m).

Convallaria majalis (lily-of-the-valley)* – a creeping rhizomatous plant with pairs of elliptic leaves and one-sided spikes of scented waxy white flowers in spring; 'Fortin's Giant' – large flowers; 'Hardwick Hall' – large broad leaves with a fine yellow margin and large flowers; 'Prolificans' (syn. 'Plena') – double flowers like little knots; 'Variegata' – gold-striped leaves. Moist humus-rich soil in partial shade; 8 in (20 cm). ✂

Cimifuga simplex

Crambe cordifolia* – massive dark green leaves and towering panicles of tiny white flowers in summer. Rich sandy soil, preferably slightly alkaline, in sun; 6 ft (1.8 m).

Crambe koktebelica* – a smaller and earlier version of the above. Cultivation as above; 3–4 ft (90 cm–1.2 m).

Crambe maritima (sea kale) – bold, waxy blue-green leaves and stout stems of cream flowers in early summer. Cultivation as above; 2 ft (60 cm).

Delphinium 'Dwarf Snow White' – pure white and quickly raised from seed; 2½–3 ft (75–90 cm); 'Galahad'* – a stately large-flowered Pacific hybrid with pure white flowers; 4–5 ft (1.2–1.5 m); 'Green Expectations' – a New Century hybrid with greenish-white flowers; 4–5 ft (1.2–1.5 m); 'Moerheimii' – a smaller, more branched Belladonna hybrid; 3–4 ft (90 cm–1.2 m). Well-drained, moisture-retentive soil in sun. ✂

Dianthus caryophyllus see under half-hardy annuals, page 100.

Dianthus x *hybridus* (old-fashioned pink) 'Mrs Sinkins'* – double white with a delicious scent; 'Musgrave's Pink' (syn. 'Charles Musgrave') –

single white with a green eye and fine scent; 'Whiteladies' – similar but rather more vigorous; summer-flowering. Well-drained, slightly alkaline soil in full sun; 6–12 in (15–30 cm). ✂

Dianthus x *allwoodii* (modern pink) – quicker growing and more floriferous than old-fashioned pinks; 'Haytor'* – perfectly shaped double white with a strong scent. Cultivation as above, 12–14 in (30–35 cm). ✂

Dianthus deltoides (maiden pink) 'Albus' – a short-lived perennial which sows itself (varieties come reasonably true from seed if grown alone); masses of miniature unscented, single white pinks over fine, compact foliage throughout most of the summer. Cultivation as above; 6–9 in (15–23 cm).

Dianthus 'Nyewood's Cream' – a miniature hybrid with dense tufted growth and tiny single creamy-white flowers in summer. Cultivation as above; 4–5 in (10–12.5 cm).

Dianthus 'Wisp' – a miniature hybrid with purple-eyed single white flowers. Cultivation as above; up to 6 in (15 cm).

Dicentra cucullaria – graceful fern-like foliage and 1 in (2.5 cm) fleshy angular heart-shaped flowers in late spring. Moist humus-rich soil in a sheltered semi-shady position; 6 in (15 cm).

Dicentra eximia 'Alba'* – grey-green fern-like foliage and clusters of fleshy white narrowly heart-shaped flowers from late spring to the end of summer. Moisture-retentive humus-rich soil in semi-shade (but tolerates drier conditions than *D. formosa*); 8–12 in (20–30 cm).

Dicentra formosa 'Alba' – similar to the above but with bright green ferny leaves and a shorter flowering period in early summer. Cultivation as above; 12 in (30 cm).

Dicentra spectabilis (bleeding heart) 'Alba'* – finely dissected leaves and arching stems from which dangle pure white heart-shaped lockets in late spring and early summer. Cultivation as above, with shelter from spring frosts and cold winds; 2–2½ ft (60–75 cm).

Dictamnus albus (burning bush)* – compound, strongly aromatic foliage and spikes of long-stamened white flowers in summer. Well-drained soil, in full sun; 2–2½ ft (60–75 cm).

Dodecatheon meadia (shooting star) 'Alba' – rosettes of smooth primula-like leaves and upright stems of small cyclamen-like flowers with pointed 'noses' and tightly reflexed petals. Moist humus-rich soil in partial shade; 10–18 in (25–45 cm).

Echinacea purpurea (purple coneflower) 'White Lustre' and 'White Swan' – large honey-scented daisies with prominent dark golden centres and drooping white ray petals; quick-flowering from seed and may be treated as an annual. Well-drained soil in sun; 18–24 in (45–60 cm). ✂

Echinops (globe thistle) 'Nivalis' – thistle-like leaves and spherical prickly-looking heads of white flowers in summer; dries well. Ordinary soil in sun; 3–4 ft (90 cm–1.2 m). ✂

Epilobium angustifolium (rosebay willowherb) 'Album'* – a truly magnificent 'weed', less invasive than the species, with tall spires of translucent white flowers, backed by green sepals. Ordinary soil in sun; 3–5 ft (90 cm–1.5 m).

Epilobium glabellum – a small New Zealand willowherb with a profusion of creamy-white flowers in summer; good at the front of borders. Cultivation as above; 9 in (23 cm).

Epilobium hirsutum (codlins-and-cream; great willowherb) 'Album' – downy lanceolate leaves and four-petalled white flowers in summer; best in the wild garden. Damp to wet soil in sun or partial shade; 4–5 ft (1.2–1.5 m).

Epimedium grandiflorum (barrenwort) 'White Queen' – attractive toothed foliage, brown in spring, accompanied by long-spurred purple-tinted white flowers. Moisture-retentive humus-rich soil in partial shade; 10 in (25 cm).

Epimedium x *youngianum* 'Niveum'* – similar to the above but with sprays of starry ivory flowers. Cultivation as above; 8–10 in (20–25 cm).

Eremurus elwesii (foxtail lily) 'Albus'* – spectacular giant white spires for the back of the border. Plant the spidery roots 6 in (15 cm) deep in a sunny position which avoids early morning sun, mulching well each spring; 6 ft (1.8 m). ✂

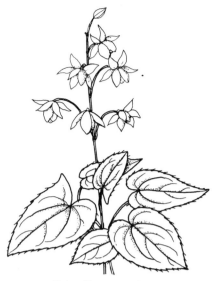

Epimedium youngianum

Eremurus himalaicus – a smaller foxtail lily with orange-anthered white flowers in the characteristic blunt spire in late spring and early summer. Cultivation as above; 2–4 ft (60 cm–1.2 m). ✂

Erinus alpinus 'Albus' – a neat little plant with evergreen toothed leaves and numerous tiny white flowers from early summer; takes to walls, paving cracks and gravel drives. Well-drained soil in sun; 3 in (7.5 cm).

Eriophorum angustifolium (cotton grass) – an invasive bog-dweller with keeled leaves and cotton-wool seedheads in summer. Wet acid soil in sun; 8–15 in (20–38 cm). ✂

Eryngium eburneum – a sea holly with a rosette of toothed, strap-shaped leaves and tall stems of long-lasting white flowerheads in summer. Well-drained soil in sun; 4 ft (1.2 m). ✂

Filipendula camschatica – a giant meadow-sweet with palmate leaves and 6–8 in (15–20 cm) panicles of fragrant white flowers in summer. Moist soil in sun or partial shade; 6 ft (1.8 cm).

Filipendula hexapetala see *F. vulgaris*.

Filipendula ulmaria (meadowsweet) – fragrant heads of cream flowers over dark, divided foliage in summer; 'Aurea' – the golden-leaved form; scorches in sun; 'Variegata' – dark green leaves slashed with yellow. Moist soil in sun or partial shade; 18 in (45 cm).

Filipendula vulgaris (syn. *F. hexapetala*) (dropwort) – deeply cut dark green leaves and large heads of tiny pink-backed cream flowers in summer; 2–3 ft (60–90 cm); 'Plena' ('Flore Pleno') – the smaller, double-flowered form; 18–24 in (45–60 cm). Moist well-drained soil, preferably alkaline, in sun.

Fragaria 'Variegata' – dark green tripartite leaves, boldy splashed with cream and small flowers with white petals and yellow centres, which unfortunately are not followed by strawberries as this form is sterile; spreads by runners; well-drained soil in sun or partial shade; 6 in (15 cm).

Fragaria vesca (wild strawberry) – dark green, toothed tripartite leaves and yellow-centred white flowers, followed by sweet aromatic red fruits; 'Flore Pleno' – double flowers; 'Fructo Alba' – white fruits. Ordinary soil in sun or shade; tolerates dry shade; spreads by runners; 6–8 in (15–20 cm).

Galega hartlandii 'Candida'* – divided leaves and numerous dense spikes of small white pea flowers in summer. Well-drained soil in sun or partial shade; 3 ft (90 cm).

Galega officinalis (goats' rue) 'Alba'* – light green divided leaves and spikes of small white pea flowers in summer. Well-drained soil in sun or partial shade; very similar to the above but more lax; 3 ft (90 cm).

Galium odoratum (syn. *Asperula odorata*) (woodruff)* – whorls of small narrow leaves topped by clusters of tiny fragrant white stars in spring; excellent deciduous ground-cover in damp shade; 9 in (23 cm).

*Gaura lindheimeri** – a Texan willowherb with narrow little leaves and open spires of the prettiest white flowers in late summer and autumn; perennial but often grown as an annual. Well-drained soil in full sun; 3 ft (90 cm).

Gentiana asclepiadea (willow gentian) 'Alba'* – rather like a Solomon's seal in growth, but with upward-facing white gentians in late summer. Damp soil in partial shade; 15–20 in (38–50 cm).

Gentian sino-ornata 'Mary Lyle' – a white-flowered form of the best autumn-flowering gentian. Moisture-retentive lime-free soil in sun; 4 in (10 cm).

Geranium – hardy geraniums are easy reliable border plants which tolerate a wide range of conditions. Some make excellent ground-cover in shade. Most are happy in ordinary soil in sun or partial shade.

Geranium x *cantabrigiense* – a hybrid between *G. macrorrhizum* and *G. dalmaticum* with a creeping habit, palmately lobed aromatic leaves and masses of short-stemmed flowers in early summer; 'Biokovo' – a Yugoslavian clone with pink-tinged white flowers; good ground-cover; 8 in (20 cm).

Geranium clarkei 'Kashmir White'* – (syn. *G. pratense* 'Kashmir White'; *G. rectum* 'Album') – a Nepalese form with large lilac-veined white flowers over finely cut foliage in summer; 2 ft (60 cm).

Geranium macrorrhizum 'Album'* – aromatic, slightly sticky foliage and pink-stamened white flowers in early summer; excellent for ground-cover and edging paths and borders; 10–12 in (25–30 cm).

Geranium phaeum (mourning widow) 'Album'* – a beautiful and easy plant for difficult damp or dry shade; large palmate leaves and clusters of small white flowers in early summer; 2 ft (60 cm).

Geranium pratense (meadow cranesbill) – long-stalked deeply divided and lobed leaves with large veined flowers in midsummer; *albiflora* – the white-flowered form; 'Kashmir White' see *G. clarkei* 'Kashmir White'; 'Plenum Album'* – small double white flowers. Adaptable, but prefers alkaline soil in sun; 2 ft (60 cm).

Geranium rectum 'Album' see *G. clarkei* 'Kashmir White'.

*Geranium renardii** – soft grey-green lobed leaves and violet-veined white flowers with notched petals; best in a sunny spot; 9 in (23 cm).

Geranium renardii

Geranium sanguineum (bloody cranesbill) 'Album' – a compact mound of finely cut leaves, studded with white flowers in summer; good for the edges of paths; 6–10 in (15–25 cm).

Geranium sylvaticum (wood cranesbill) 'Album'* – elegantly cut leaves and small white flowers in late spring and early summer; 2 ft (60 cm).

Geum rivale (water avens) 'Album' – an unassuming but charming summer-flowering plant for damp shady places; reddish stalks and pendant bell-shaped flowers with large green calcyces from which peep white petals; 12 in (30 cm).

*Gillenia trifoliata** – a graceful plant with reddish-brown stems and clover-like leaves, aflutter with dainty white flowers in early summer. Moisture-retentive soil in partial shade; 3 ft (90 cm).

Gypsophila paniculata (baby's breath) – grey-green grassy leaves and a cloud of little white flowers in summer; 'Bristol Fairy' – double white; 'Compacta Plena' – a double white variety which is only half the usual height. Neutral to alkaline soil in sun; 3 ft (90 cm). ✂

Helleborus niger (Christmas rose) – fine white flowers (sometimes tinged pink) open widely above the emerging dark green leathery leaves

in winter; 'Potter's Wheel' – glossy foliage and flowers up to 5 in (12.5 cm) across; resents disturbance; needs protection from bad weather and slugs. Moist humus-rich soil in sun or partial shade; 10–15 in (25–38 cm). ✂

Helleborus orientalis (Lenten rose) – white clones of this variable species are worth obtaining from nurseries that label the different colours (which may be anything from purple to pink and greenish-white); large, almost pendant flowers, several to a stem, in winter and early spring. Moist well-drained soil in partial shade; 15 in (38 cm). ✂

Hepatica nobilis (syn. *H. triloba*) – semi-evergreen three-lobed leaves and posies of fragile anemones in late winter and spring; 'Alba' – white; 'Alba Plena' – the rare double white form. Moist humus-rich soil in partial shade; dislikes disturbance, deep planting, and polluted air; 4 in (10 cm).

Hepatica triloba see *Hepatica nobilis*.

Heuchera sanguinea (coral flower) 'Pearl Drops' – evergreen geranium-like leaves and wiry stems of tiny white bells in summer; good ground-cover and very pretty *en masse*. Well-drained soil in sun or partial shade; 12–18 in (30–45 cm). ✂

Hosta albomarginata see *H. sieboldii*.

Hosta 'Golden Medallion' – rounded puckered gold leaves and white flowers in early summer; good ground-cover. Moist humus-rich soil in partial shade; 18 in (45 cm). ✂

Hosta plantaginea – a plantain lily with large heart-shaped leaves which retain their fresh yellowish-green until late in the growing season; fragrant white flowers in autumn if growing conditions are to its liking; 'Grandiflora' – larger flowers. Rich moist soil in a sheltered position; 18–24 in (45–60 cm). ✂

Hosta 'Royal Standard'* – a hybrid which has *H. plantaginea* as one parent; large light green heart-shaped leaves and sturdy, freely produced stems of scented white flowers in late summer. Cultivation as above; 2½–3 ft (75–90 cm). ✂

Hosta sieboldii 'Alba'* – narrow white-margined leaves and spikes of white trumpet-shaped flowers in midsummer; guaranteed (as

are all hostas) to delight gardeners, flower arrangers – and slugs. Rich moist soil in partial shade; 18 in (45 cm). ✂

Hosta 'Snowden' – a giant plantain lily with graceful grey-green foliage complemented by white flowers in late summer. Cultivation as above; 4 ft (1.2 m). ✂

Houstonia caerulea (bluets) 'Alba' – a miniature chickweed-like species for the rock garden or containers; bright green mats studded with four-petalled white stars in spring and summer. Moist lime-free soil in shade; 2 in (5 cm).

Houttuynia cordata – creeping erect stems of small purplish-green heart-shaped leaves which have a powerful orange-like smell; small cone-shaped flowers with showy white bracts in summer; 'Flore Pleno' – the double-flowered form; 'Chameleon' (syn. 'Variegata') – single flowers and less vigorous but with sensational leaves vividly marbled in green, yellow, red and bronze; spreads quickly in damp or wet soil in sun or partial shade; 10–18 in (25–45 cm).

Iberis sempervirens (perennial candytuft) – a reliable edging and rockery plant with narrow dark evergreen foliage and rounded heads of tiny brilliant white flowers in spring; 'Little Gem' – a smaller, more compact form; 'Plena' – double-flowered; 'Snowflake' – a dense spreading form which reaches 2 ft (60 cm) across and flowers in early spring. Ordinary well-drained soil in sun; 6–10 in (15–25 cm).

Iris albicans (syn. *I. florentina* var. *albicans*) – the Moslem flowers of the dead; broad glaucous foliage and yellow-bearded pure white flowers in early summer. Well-drained soil in sun; 2 ft (60 cm). ✂

Iris 'Cliffs of Dover'* – a tall vigorous bearded hybrid with large ruffled creamy-white flowers in early summer. Well-drained soil in sun; 3½ ft (1.05 m). ✂

Iris florentina var. *albicans* see *Iris albicans*.

Iris 'Frost and Flame'* – a tall bearded hybrid with crisp white flowers and contrasting orange-red beards. Cultivation and height as above. ✂

Iris 'Green Spot' – a dwarf bearded hybrid, popular since its creation in 1951, with distinctive green-marked horizontal ivory falls; spring-flowering. Cultivation as above; 9 in (23 cm).

Iris hoogiana 'Alba' – grey-green leaves and scented satin milk-white flowers in late spring. Hardy only in a dry warm sheltered position; 2½ ft (75 cm).

Iris japonica 'Ledger's Variety' – the hardiest form; evergreen fans of glossy leaves and small frilly flowers in bluish-white, marked blue and orange. Moist humus-rich soil in a warm sunny position with some protection in harsh winter weather; 12–18 in (30–45 cm).

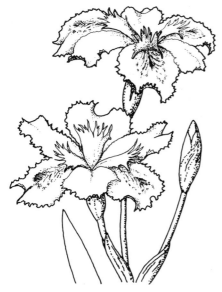

Iris japonica

Iris kaempferi (Japanese iris) 'Alba' – deciduous foliage and branched stems of three or four large flowers with broad drooping white petals in summer; 'Moonlight Waves' – white with a green centre; dislikes lime. Moist, rich soil in sun; 18 in (45 cm). ✂

Iris 'Kanchenjunga' – a large bearded hybrid with pure white flowers. Well-drained soil in sun; 3½ ft (1.05 m). ✂

Iris laevigata – light green deciduous foliage and stems of three flowers in summer; 'Alba' – flat white flowers; 'Snowdrift' – a good white variety. Damp to wet soil or shallow water; 18–24 in (45–60 cm). ✂

Iris 'Langport Snow' – an intermediate iris with pure white flowers in late spring. Well-drained soil in sun; 16 in (40 cm). ✂

Iris 'Lilliwhite'* – a dwarf bearded hybrid with pure white flowers. Well-drained soil in sun; 12 in (30 cm).

Iris sanguinea 'Alba' – dense narrow foliage and beautifully shaped white flowers in summer. Ordinary soil in sun or partial shade; 3 ft (90 cm). ✂

Iris sibirica – rush-like leaves and branched stems of smallish flowers in early summer. There are many white forms, including 'Alba', 'Fourfold White', 'Snow Princess', 'Snow Queen', 'White Queen', 'White Swirl', 'Wisley White' and 'Lime Heart' (the latter being green-centred). Damp soil in sun; 2½ ft (75 cm). ✂

Iris stylosa see *I. unguicularis*.

Iris tectorum (roof iris) 'Alba' – a Chinese species with frilly white flowers in early summer. Moist humus-rich soil in partial shade; 12–18 in (30–45 cm).

Iris 'The Bride' – a dwarf spring-flowering variety with primrose-bearded white flowers. Well-drained soil in sun; 8 in (20 cm).

Iris unguicularis (syn. *I. stylosa*) 'Alba' – the rare white form of the winter-flowering Algerian iris; grassy leaves and perfect white rises with yellow markings on the falls; slow to establish. Poor dry soil in full sun, between paving stones or against a wall in cold areas; 10–12 in (25–30 cm). ✂

Iris 'White City' – a hybrid bearded iris with smallish blue-white flowers in early summer. Well-drained soil in sun; 3½ ft (1.05 m). ✂

Kniphofia (red-hot poker) 'Little Maid' – a white-hot poker with tidy linear foliage and long spikes of narrow tubular flowers which are green in bud and open to ivory; flowers all summer and into autumn. Well-drained soil in sun; 2 ft (60 cm).

Lamium maculatum (dead nettle) 'Album' – ground-covering silver-striped leaves and white flowers in late spring and early summer; 'Beedham's White' – a non-rampant white-flowered golden form which needs rich soil

and cool shade; 'White Nancy'* – silver foliage and white flowers, ideal for the front of silver borders; ordinary soil in sun or shade; 5–6 in (12.5–15 cm).

Lamium orvala (giant dead nettle) 'Album' – a well-behaved herbaceous dead nettle, tolerant of heavy shade. Moisture-retentive soil; 2 ft (60 cm).

Leontopodium alpinum (edelweiss) – narrow grey-green leaves and unusual blooms with woolly off-white bracts surrounding tiny yellowish groundsel-like flowers, symbolic of alpine idylls. Well-drained soil in sun on the rock garden or in stone troughs; 6–8 in (15–20 cm).

Leucanthemum maximum (syn. *Chrysanthemum maximum* (Shasta daisy) – lanceolate toothed dark green leaves and strong stems of large yellow-centred white daisies; 'Beauté Nivelloise'* – narrow shaggy petals; 'Esther Read' – fully double white pompoms; 'H. Siebert' – huge single white with frilled petals; 'Horace Read' – similar but more creamy-white; 'Phyllis Smith' – finely cut ray petals; 'Snowcap'* – perfect single daisies, dwarf and floriferous; 'T. E. Killin' – semi-double with greenish-yellow centres, beloved of flower arrangers; 'Wirral Pride' – semi-double, with a circle of close-cropped ray petals around the yellow centre; 'Wirral Supreme'* – double white; well-drained soil in sun; 18–36 in (45–90 cm). ✂

Leucanthemella serotina see *Chrysanthemum uliginosum*.

Leucanthemum vulgare (ox-eye daisy) – long-stemmed single daisies. Easily grown in most soils; naturalises in grassland; 2 ft (60 cm). ✂

Liatris scariosa 'Alba Magnifica' – similar to the above but taller and with less dense spikes. Ordinary soil in sun; up to 5 ft (1.5 m).

Liatris spicata (blazing star; gayfeather) 'Alba' – an excellent border plant with grassy leaves and creamy-white pokers which open from the top downwards in late summer; 'Floristan White' – pure white. Fairly tolerant but happiest in rich damp soil; 2–2½ ft (60–75 cm). ✂

Libertia formosa – an iris relative from Chile, with graceful evergreen foliage and tall panicles of small white flowers in summer, fol-

Libertia formosa

lowed by attractive brown seedheads. Well-drained soil enriched with sand, peat and leafmould, in a sunny position; 2–3 ft (60–90cm).

Libertia grandiflora – a larger flowered species from New Zealand which needs a warm situation; three-petalled flowers in summer, followed by orange fruits. Cultivation and height as above.

Linum perenne (flax) 'White Diamond' – a lovely plant for the silver garden, with a compact habit, narrow grey-green leaves and wiry stems carrying a succession of simple pure white flowers; a short-lived perennial but easily raised from seed. Ordinary soil in full sun; 12in (30cm).

Liriope spicata 'Alba' – clumps of grassy, dark green leaves and stiff, slender spikes of white flowers in autumn; a useful evergreen for the front of borders; tolerates drought but dislikes alkaline soil. Light sandy soil in sun or partial shade; 8–10in (20–30cm).

Lobelia syphilitica 'Alba' – a short-lived perennial from North America with slender light green leaves and tall spikes of white flowers in summer. Rich damp soil in sun or partial shade; 2 ft (60cm).

Lunaria rediviva (honesty) a short-lived perennial honesty with dark green serrated leaves and scented springtime flowers in palest lilac-white, followed by elliptic papery seed cases. Light soil in partial shade; up to 3½ft (1.05m).

Lupinus (lupin) – classic summer border plants with soft green digitate leaves and dense spires of peppery-smelling pea flowers; 'Blushing Bride' – blush white; 'Noble Maiden'* cream buds opening to ivory. Best in neutral to acid soils in sun or partial shade; 3 ft (90cm) or more.

Lychnis alpina (syn. *Viscaria alpina*) 'Alba' – an alpine campion with tufts of narrow dark green leaves and dense 2in (5cm) heads of small five-petalled flowers in early summer. Ordinary soil in sun; 4in (10cm).

Lychnis chalcedonica (Maltese cross) 'Alba' – large slightly rounded heads of small white flowers, each shaped like a Maltese cross, in summer. Ordinary soil in sun; 2–3 ft (60–90cm).

Lychnis coronaria 'Alba'* – an indispensable but short-lived perennial for the white garden; grey foliage and white campion flowers in summer. Ordinary soil in sun; 2ft (60cm).

Lychnis flos-cuculi (ragged robin) 'Albiflora' – very lovely fimbriate flowers in early summer, with petals divided into four narrow lobes. Damp to marshy soil in sun or partial shade; naturalises well in poorly drained grassland; 10–18in (25–45cm).

Lychnis flos-jovis 'Alba' – silvery foliage and white campion flowers in early summer. Ordinary soil in sun; 6–10in (15–25cm).

Lychnis viscaria (sticky catchfly) 'Alba' – whorls of white campion flowers in early summer on leafy stems which have a sticky patch below each leaf node. Ordinary to dry stony soil in sun; 10–12in (25–30cm).

Lysichiton camtschatcensis (eastern skunk cabbage)* – an architectural plant for boggy places; huge white spathes and greenish spadices which emerge directly from the mud in early spring, followed by luxuriant oval leaves. Rich wet soil in sun or partial shade; 3 ft (90cm).

Lysimachia clethroides* an oriental loosestrife which spreads enthusiastically into bold clumps of tall leafy stems, topped in late summer by compact, crook-shaped spikes of starry white flowers. Moisture–retentive soil in sun or partial shade; 3 ft (90cm).

Lysimachia ephemerum – a well-behaved European species with narrow grey–green leaves ascending slender stems and fine long spires of white flowers in summer. Cultivation as above; 3 ft (90cm).

Maianthemum bifolium (May lily) – a miniature carpeter for cool shady places; shiny heart-shaped leaves, two to a stem, and spikes of tiny fragrant white flowers in early summer, followed by red berries; good ground-cover under and around conifers on neutral to alkaline soil; 6in (15cm).

Malva moschata (musk mallow) 'Alba'* – an easy and excellent border plant with attractive divided leaves and the loveliest white mallow flowers over a long period in summer; comes true from seed and sows itself modestly. Ordinary soil in sun or partial shade; 18in (45cm),

Matricaria inodora (syn. *M. maritima*) (scentless mayweed; sea mayweed; corn feverfew) 'Bridal Robe' – the semi-double form of a familiar wild flower. Ordinary soil, including dry and sandy conditions, in sun; 2ft (60cm). ✂

Mazus reptans 'Albus' – a tiny Himalayan creeping perennial which forms mats of light green lanceolate leaves which in summer are studded with relatively large squat lobelia-like flowers. Moist soil in sun; 1 in (2.5cm).

Meconopsis betonicifolia (syn. *M. baileyii*) 'Alba' – the rare white form of the Himalayan blue poppy and a beautiful plant in its own right; a short-lived perennial if the flowers are removed the first year, though otherwise usually dying after flowering. Rich moist fibrous lime-free soil in light shade; 3 ft (90cm).

Meconopsis superba – a rosette of greyish hairy leaves and ivory poppies in early summer. Cultivation and characteristics as above; 3 ft (90cm).

Meum athamanticum (spignel) – a more delicate version of cow parsley with very finely cut aromatic leaves and umbels of tiny off-white flowers in summer. Ordinary soil in sun; 2 ft (60 cm).

Monarda fistulosa (wild bergamot) 'Snow Maiden' – leaves with the scent of Earl Grey tea and whorls of large white claw-shaped flowers in summer; tolerates drier conditions than varieties of *M. didyma*. Humus-rich soil in sun or partial shade; 3 ft (90 cm).

Myrrhis odorata (sweet cicely) – soft ferny aniseed-scented foliage and umbels of white flowers in early summer followed by long shiny black seeds. Moisture-retentive soil in sun or shade; 2–3 ft (60–90 cm).

Nierembergia repens (syn. *N. rivularis*) – a tiny Argentinian creeping species with mats of oval leaves and disproportionately large pure white cup-shaped flowers in summer. Moisture-retentive soil in a sheltered, sunny position; 2 in (5 cm).

Nierembergia repens

Nierembergia rivularis see *Nierembergia repens.*

Nymphaea alba* – the classic white European water lily; almost round, light green lily pads and fragrant flowers with yellow stamens in summer, still or slow-moving water up to 3 ft (90 cm) deep.

Nymphaea 'Albatross' – a hybrid water lily with bright green leaves and pure white semi-double flowers. Needs fairly shallow water no more than 18 in (45 cm) deep.

Nymphaea candida – a free-flowering small-growing species suitable for pools and sinks with a depth of 4–9 in (10–23 cm).

Nymphaea 'Caroliniana Nivea' – a large-flowered, very fragrant variety for small pools 6–12 in (15–30 cm) deep.

Nymphaea 'Gladstoniana'* – a breathtaking variety, probably the largest, with white flowers 8–10 in (20–50 cm) across. Needs a water depth of 4–6 ft (1.2–1.8 m).

Nymphaea x *marliacea* 'Albida'* – a reliable easy medium-sized hybrid with an abundance of light green leaves and pure white flowers. Best in 6–18 in (15–45 cm) of water.

Nymphaea odorata – a North American species with bright green leaves and scented flowers which is the parent of the majority of hardy hybrids; represented in cultivation by: 'Alba' – an easy free-flowering scented variety for pools 6–18 in (15–45 cm) deep; 'Minor' – a smaller form with dark leaves and relatively few slightly scented 2–3 in (5–7.5 cm) flowers, which can be grown in small pools and containers where the water is 6–12 in (15–30 cm) deep.

Omphalodes verna (blue-eyed Mary) 'Alba' – a delightful ground-covering plant with light green oval leaves and sprigs of chalk white flowers, resembling forget-me-nots, in spring; spreads by runners. Moist soil, enriched with peat, in shade; 4 in (10 cm).

Osteospermum see under tender perennials, page 131.

Pachyphragma macrophylla (syn. *Cardamine asarifolia*) – an Italian lady's smock with racemes of white cress-like flowers in early spring, followed by bold rounded leaves which resemble those of *Asarum*; good ground-cover in shade. Ordinary moisture-retentive soil; 12 in (30 cm).

Paeonia (peony) – herbaceous peonies should be planted shallowly in rich well-drained soil and mulched annually with compost or well-rotted manure, avoiding digging or forking around the root space. They do well in sun or partial shade but sites which admit early morning sun to frosted plants should be avoided. The commonest problem with peonies is for the flower buds to remain small, hard and unopening. It can be caused by rapid warming after frost, over-deep planting, root disturbance, inadequate feeding or excessive dryness.

Paeonia albiflora see *P. lactiflora.*

Paeonia emodi – an Indian species with single white flowers in summer; 18–30 in (54–75 cm).

Paeonia lactiflora (syn. *P. albiflora*) – the parent of many hybrids; handsome divided leaves and scented single white flowers in early summer; 'Alice Harding' – large double ivory flowers flushed with gold; 'Barrymore' – a huge single white, blush on opening, with a centre of yellow petaloids; 'Blush Queen' – a large creamy white double; 'Charles White' – early variety with long-lasting fragrant flowers; reaches 4 ft (1.2 m); 'Couronne d'Or' – a creamy white double; 'Duchesse de Nemours'* – a hybrid dating from 1856 (but still one of the best) with medium-sized, very fragrant white flowers, shading to yellow in the centre and produced over a long period in early summer; 'Festiva Maxima' – a magnificent Victorian hybrid with large fragrant, loosely double flowers whose white petals are flecked with crimson; 'Krinkled White' – a vigorous early flowering variety with huge single petals, attractively crinkled and faintly flushed pink at the edges; 'Le Cygne' – pure white double; 'White Wings' – a fine single variety with a prominent boss of yellow stamens; 'Whitleyi Major'* (syn. 'The Bride') – large flowers, very like the species and probably the nearest that is widely available. Unless stated otherwise, 2–3 ft (60–90 cm). ✕

Paeonia obovata 'Alba' – single silky white flowers in late spring; 'Grandiflora' – large cream flowers in late spring and early summer; up to 2 ft (60 cm).

Paeonia officinalis 'Alba Plena'* – the double blush white form of the common European peony which flowers in early summer; 18–24 in (45–60 cm). ✕

Papaver orientale (oriental poppy) 'Black and White'* – a stunning poppy with large crinkled white flowers, dramatically blotched with black at the base of each petal; 'Perry's White' – blush-white, sometimes with a smudge of purplish-black in the centre. Ordinary soil in sun; 2–2½ ft (60–75 cm).

Paradisea liliastrum (St Bruno's lily) – an alpine plant from southern Europe with tall stems of scented tubular flowers in early summer. Rich sandy soil in partial shade; 12–24 in (30–60 cm).

Parnassia palustris (grass of Parnassus) – a delightful little plant for damp places; heart-shaped leaves and single, beautifully veined white flowers on long stalks in summer. Moist to wet soil, including grassland, in sun or shade; 6–9 in (15–23 cm).

Parnassia palustris

Penstemon x *gloxinioides* – hybrids derived from *P. hartwegii* which bear spikes of 2 in (5 cm) dropping tubular flowers in summer; 'White Bedder' – one of the few white penstemons. Well-drained soil in sun; 2 ft (60 cm).

Penstemon rupicola 'Albus' – a prostrate shrubby perennial with small ovate leaves and clusters of 1 in (2.5 cm) flowers in early summer. Cultivation as above; 4–6 in (10–15 cm)

Phlox adsurgens 'Alba' – a prostrate species, ideal for rock gardens and containers, which forms mats of small shiny oval leaves with generous clusters of five-petalled white flowers in early summer. Well-drained peaty soil in sun or partial shade; 6–10 in (15–25 cm).

Phlox douglasii 'May Snow', 'Snow Queen' and 'White Drift' – prostrate mat-forming plants, smothered in white flowers in late spring; 'Iceberg' – neat mats of blue-tinged white flowers; 2 in (5 cm); reliable and easy for rock gardens, walls and containers; more tolerant of dry conditions than *P. adsurgens*. Ordinary soil in sun; 3–4 in (7.5–10 cm).

Phlox maculata – maroon-spotted stems, lanceolate leaves and narrow tapering 6 in (15 cm) heads of scented flowers in late summer; 'Miss Lingard'* – brilliant white; 'Omega' – violet-tinged white flowers with a pink eye. Rich moisture-retentive soil in sun or partial shade; 2–3 ft (60–90 cm).

Phlox nivalis – a compact plant for the rock garden with variable flowers in summer; 'Gladwyn' – white with an orangey-yellow pinpoint centre, 2–4 in (5–10 cm); 'Nivalis' – tight cushions and neat white flowers, 1 in (2.5 cm). Well-drained soil in sun.

Phlox paniculata 'Alba'* – the white form of the summer-flowering border phlox; 'Blue Ice' – pink buds opening to bluish-white with a pink eye; 'Fujiyama'* – a strong variety with large cylindrical heads of pure white flowers; 'Mother of Pearl' – blush-white; 'White Admiral' – perfectly shaped domes of white flowers. Rich moisture-retentive soil in sun or partial shade; 3½–4 ft (75 cm–1.2 m).

Phlox subulata (moss phlox) 'White Delight' – a mat-forming species for the rockery and similar situations; light green linear leaves and dense masses of white flowers in spring. Ordinary soil in sun; 2–4 in (5–10 cm).

Physostegia virginiana (obedient plant) 'Alba' and 'Summer Snow'* – valuable autumn-flowering border plants with white mini-snapdragon flowers. Moisture-retentive soil in sun or partial shade; 2–3 ft (60–90 cm).

Platycodon grandiflorus (balloon flower) 'Albus'* – a beautiful but fragile plant with grey-green ovate leaves and inflated buds which open into giant campanula-like flowers in summer; 'Mariesii Albus' – big white flowers on small plants, 9–12 in (23–30 cm); 'Snowflakes' – semi-double; resents disturbance and can be difficult to establish; disappears completely in winter and reappears rather late in spring, so mark the spot. Ordinary soil in sun; 12–24 in (30–60 cm).

Podophyllum emodi see *Podophyllum hexandrum*.

Podophyllum hexandrum (syn. *P. emodi*) (Himalayan May apple) – a choice though poisonous plant for rich moist soil in shade; fragile crinkled white flowers in spring above newly unfolding leaves which are brown–marbled and decoratively lobed and toothed; large egg-shaped red fruits; 12 in (30 cm).

Polemonium caeruleum (Jacob's ladder) 'Alba' – neat pinnate (ladder-like) leaves and erect branched clusters of five-petalled flowers over a long period in summer. Ordinary soil in sun or partial shade; 2 ft (60 cm).

Polemonium foliosissimum 'Alba' – a Rocky Mountain Jacob's ladder, very similar to *P. caeruleum* but taller and later. Cultivation as above; 3 ft (90 cm).

Polygonatum commutatum see *Polygonatum giganteum*.

Polygonatum falcatum (Solomon's seal) – reddish stems and dangling white bells in spring; 'Variegatum' – white-margined leaves. Moisture-retentive soil, enriched with peat and leafmould, in shade; 2½ ft (75 cm). ✂

Polygonatum giganteum (syn. *P. commutatum*) (giant Solomon's seal) – similar to above, but much larger. Rich moist soil in shade; averages 3½ ft (105 cm) but may reach 7 ft (2.1 m) in good conditions.

Polygonatum x *hybridum* (Solomon's seal)* – often referred to as *P. multiflorum*, the wild species which is one of its parents (the other being *P. odoratum*); a tough reliable border plant with distinctive arched stems clad in stiff rows of elliptic leaves from which dangle white, green-tipped bells in late spring, followed by black berries; 'Variegatum' – smaller than the variegated form of *P. falcatum*, with leaves striped and margined white. Ordinary soil in shade; 2–4 ft (60–90 cm). ✂

Polygonatum odoratum (angular Solomon's seal) 'Gilt Edge' – a slow-growing, small Solomon's seal with reddish angled stems, gold-edged leaves and scented bells which, unlike those of most other species, are not eggtimer-shaped when in bud, 16 in (40 cm); 'Variegatum' – similar but white-variegated and rather larger, 2½ ft (75 cm). Rich moist soil, preferably alkaline, in shade.

Polygonatum verticillatum (whorled Solomon's seal) – a distinctive species with erect stems, narrow leaves in whorls, and very small green-tipped white bells in pendant clusters in summer, followed by red berries. Rich moist soil in shade; 3½ ft (1.05 m).

Polygonum amplexicaule (knotweed) 'Album' – dense mounds of neat foliage topped by longlasting slender white spikes in summer; 4 ft (1.2 m).

Potentilla alba – ground-covering mats of grey-green palmate leaves and orange-eyed white flowers in spring. Ordinary soil in sun or light shade; 4 in (10 cm).

Potentilla rupestris – small yellow-eyed white flowers on pink stems in early summer. Moisture-retentive soil in sun; 18 in (45 cm).

Poterium magnificum (burnet) 'Album' – greyish pinnate leaves with serrated margins and drooping white bottle-brushes in summer. Moist soil in sun or shade; 2½ ft (75 cm).

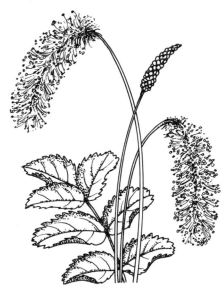

Poterium magnificum

Pratia treadwellii – a delightful miniature carpeter for a cool shady spot on the rock garden; tiny round leaves, white starry flowers and pinkish-purple berries; 2 in (5 cm).

Primula – most primulas need humus-rich moist soil and partial shade. Equally important, they need annual division or repotting. It is also a good idea to avoid planting primulas in the same place year after year as this may lead to the soil becoming infected with 'primula sickness'. Many species and varieties grown from seed are variable, in which case exceptional plants must be earmarked for vegetative propagation.

Primula alpicola (moonlight primula) – like a large cowslip with long-stalked leaves and scented, generally cream flowers but varying from violet to white if grown from seed; early summer-flowering; 'Alba' – the white form; 'Luna' – creamy-white with a touch of lemon. Humus-rich moist soil in semi-shade; 6–18 in (15–45 cm).

Primula auricula – ovate to obovate leaves, sometimes grey-green and mealy, and umbels of sturdy flowers in spring; 'Devon Cream' – a delightful double cream variety, the nearest yet to a white auricula. Well-drained humus-rich soil in sun or partial shade; 6 in (15 cm).

Primula chionantha – a tall and relatively easy summer-flowering 'Nivalid' species from Yunnan; powdery leaves, stalks dusted lime-green, and heads of delicately scented white flowers, attractively tinged with greyish-mauve. Rich moist soil in partial shade; 12–18 in (30–45 cm).

Primula denticulata (drumstick primrose) 'Alba'* – a superb form of a good-natured Himalayan species, with ball-shaped heads of white flowers in spring; thrives in most soils and conditions; 12 in (30 cm).

Primula japonica (Japanese primrose) 'Fuji'* – tall stout stems of tiered white flowers; 'Postford White' – yellow-eyed white flowers from late spring to midsummer. Moisture-retentive soil in partial shade; 18 in (45 cm).

Primula x *juliana* see *P.* x *pruhoniciana*.

Primula marginata 'Alba' – a reasonably easy alpine primula with powdery grey-green silver-edged leaves among which nestles a profusion of scented flowers in spring. Well-drained soil in shady crevices on the rock garden or in a drystone wall; also good in pots and containers; 4 in (10 cm).

Primula x *pruhoniciana* (syn. *P.* x *juliana*) – a wide variety of hybrids characterised by the dark-leaved purple-flowered 'Wanda'; 'Garryarde Sir Galahad Seedling' – bronze leaves and creamy-white flowers, 4–6 in (10 cm); 'Snow White' – a compact white-flowered variety only 2 in (5 cm) high. Moist humus-rich soil in sun or partial shade.

Primula x *pubescens* – hybrids between *P. auricula* and various other species, including

P. hirsuta; variable in size and appearance but generally with obovate leaves and dense umbels of flowers in spring; 'Bewerley White' – a vigorous variety with large umbels of creamy-white flowers; 'Harlow Car' – white, faintly blushing pink with age; 'Nivea' – white. Well-drained gritty humus-rich soil in sun or partial shade; excellent for pots in the cold greenhouse; 4–6 in (10–15 cm).

Primula sieboldii – a Japanese woodland species with light green ovate leaves, distinctly toothed at the margins, and umbels of 1 in (2.5 cm) flowers, occasionally with fringed petals, in early summer; 'Alba'* – pure white; 'Winter Dreams' – snowflake flowers. Humus-rich moist soil in dappled shade; 6–9 in (15–23 cm).

Primula x *variabilis* (polyanthus) – hybrids originally derived from the primrose (*P. vulgaris*) and the cowslip (*P. veris*); bold umbels of flowers throughout the spring; 'Lady Greer' – cream; 'McWatt's Cream' – tiny cream flowers. Moist soil in sun or partial shade; 6–9 in (15–23 cm). ✂

Primula vulgaris (primrose) 'Alba'* – the white form of the common primrose, a delightful posy of a plant which bears little resemblance to the whites among the ungainly giant hybrids which flood the garden centres around Mother's Day; 'Alba Plena'* – the incomparable double white primrose; 'Dawn Ansell'* – a double white jack-in-the-green primrose (i.e. the flowers are backed by a ruff of miniature leaves). Humus-rich moist soil in partial shade; 4–6 in (10–15 cm). ✂

Prunella grandiflora (self-heal) – slightly invasive but useful ground-cover on rock gardens and along path edges; ovate leaves and spikes of tubular flowers from late spring to autumn; 'Alba' – shows up better in shade than the purple. Moisture-retentive soil in sun or partial shade; 6 in (15 cm).

Prunella vulgaris 'Alba' – a white-flowered common self-heal. Cultivation as above; 4 in (10 cm).

Prunella webbiana – similar to the above but with broader leaves and stubbier, more showy spikes of flowers; 'Alba' and 'White Loveliness' – both pure white; good for the front of borders. Cultivation as above; 6–10 in (15–25 cm).

Pulmonaria officinalis (lungwort; Jerusalem cowslip; spotted dog) – elliptic rough-textured white-spotted leaves and clusters of small funnel-shaped flowers in spring; 'Alba' – white flowers; 'Sissinghurst White'* – an excellent white form. Moist soil in shade; 12 in (30 cm).

Pulmonaria saccharata – similar to the above species but slightly earlier and more refined; narrow ovate leaves marbled and spotted with silvery pale green; 'Alba' – white; 'Wisley White'* – a good show of white flowers. Ordinary soil in shade; 10–12 in (25–30 cm).

Pulsatilla alpina – ferny foliage, silky bluish buds and downy white anemones in early summer. Well-drained soil in sun; 12 in (30 cm).

Pulsatilla vernalis – a smaller pasque flower with furry brown buds opening to white anemones in spring. Cultivation as above but needs protection from winter wet; 6 in (15 cm).

Pulsatilla vulgaris (pasque flower) 'Alba'* – a lovely form of a most beautiful species; feathery foliage, downy large white anemone flowers, a boss of yellow stamens, and silky plumed seedheads. Well-drained alkaline soil in sun; 8–12 in (20–30 cm).

Pyrethrum roseum see *Chrysanthemum coccineum*.

Ranunculus aconitifolius (fair maids of France) – palmate leaves and small glossy white buttercups in early summer; 'Flore Pleno'* – the more usually grown double form; lovely with white-variegated hostas. Moist soil in sun or partial shade; 2 ft (60 cm).

Rodgersia pinnata 'Elegans' – handsome horse-chestnut-like leaves, often bronzed, and branched panicles of scented cream flowers in summer. Moist soil enriched with compost and peat in semi-shade; 3 ft (90 cm).

Rodgersia tabularis – a striking plant with large round leaves on centrally attached stalks and cream flowers in summer. Cultivation as above; 3 ft (90 cm).

Sagina glabra (pearlwort) – evergreen mats of moss-like foliage and minuscule white stars in summer; 'Aurea' – golden foliage, widely used by floral clock school of carpet bedding and as a substitute for grass in lawns on sandy soil. Sandy soil in sun; ½–2 in (1–5 cm).

Rodgersia pinnata

Sagittaria japonica see *S. sagittifolia* ssp. *leucopetala*.

Sagittaria sagittifolia (arrowhead) – an aquatic with ribbon-like underwater leaves, handsome arrow-shaped emerged leaves and tall stems of purple-centred three-petalled white flowers in late summer; produces stolons on which overwintering tubers develop. Suitable for ponds or streams with a depth of anything between 6 in (15 cm) and 3 ft (90 cm); 12–30 in (30–75 cm).

Sagittaria sagittifolia ssp. *leucopetala* (syn. *S. japonica*) (Japanese arrowhead) – very similar to the above but with slightly larger yellow-stamened flowers; 'Flore Pleno' – paler leaves and double stock-like flowers. Prefers shallow water, 3–6 in (7.5–15 cm) deep; 18–30 in (45–75 cm).

Salvia argentea* – a short-lived perennial with a rosette of large silky grey leaves, which elongates in summer to produce panicles of nearly white, pink-flushed hooded flowers. Well-drained soil in sun; 18–24 in (45–60 cm).

Salvia officinalis (common sage) 'Albiflora' (syn. 'Alba')* – muted sage-green foliage and spikes of white flowers in summer. Well-drained soil in sun; 12–24 in (30–60 cm).

Sanguisorba canadensis (burnet) – light green pinnate leaves and 6 in (15 cm) bottle brush heads of greenish-white flowers in late summer. Moist soil in sun or shade; 4–5 ft (1.2–1.5 m).

Saponaria officinalis (soapwort; goodbye-to-summer) 'Alba Plena' – loosely double campion flowers on sprawling stalks in late summer; long-lived and rather invasive. Ordinary soil in sun or partial shade; 2–2½ ft (60–75 cm). ✂

Saxifraga cochlearis – mounds of lime-encrusted rosettes and panicles of upward-facing white flowers in summer, 6 in (15 cm); 'Minor' – miniature rosettes which form hard silvery mounds, 4 in (10 cm). Plant vertically in gritty alkaline soil in semi-shaded rock crevice or wall cranny, or in pots.

Saxifraga cortusifolia (syn. *S. fortunei*) – scalloped leaves with reddish undersides and sprays of delicate white flowers in autumn; 'Rubrifolia' – shiny burgundy leaves with pink undersides and flower stalks; 'Wada' – dark purplish-bronze foliage. Ordinary soil in a cool, shady situation; 12–15 in (30–38 cm).

Saxifraga cotyledon – an alpine with neat lime-encrusted rosettes which erupt into long, graceful sprays of starry white flowers in summer. Cultivation as above; 2 ft (60 cm).

Saxifraga fortunei see *S. cortusifolia*.

Saxifraga moschata – cushions of moss-like foliage and small clusters of white flowers on wiry stalks in summer; 'Cloth of Gold' – yellow foliage which needs shade; 'Pearly King' – a good white form; excellent for the rock garden and sides and paths. Moisture-retentive soil in semi-shade; 3–4 in (8–10 cm).

Saxifraga oppositifolia 'Alba' – creeping mats of tiny wedge-shaped leaves and ½ in (1 cm) white cup-shaped flowers in spring. Peaty soil in a cool shady spot on the rock garden; 1 in (2.5 cm).

Saxifraga stolonifera (mother of thousands) – like a strawberry plant in habit, with rosettes of scalloped roundish leaves, silvery veins and reddish undersides, and long runners which sprout plantlets; graceful racemes of unequal-petalled white flowers in summer; 'Tricolor' – pink-edged leaves; not reliably hardy; may be grown as a trailing pot plant. Ordinary soil in semi-shade; 9–12 in (23–30 cm).

Scabiosa caucasica (scabious) – clumps of jagged and almost leafless stalks of 3 in (7.5 cm) flowers throughout the summer; 'Bressingham White' – a good white form; 'Miss Willmott' – ivory; 'Mount Cook' – large white with a green centre. Well-drained soil in sun; 18–24 in (45–60 cm). ✕

Scutellaria alpina 'Alba' – a mat-forming alpine skullcap with white hooded flowers in late summer. Well-drained soil in sun or partial shade; 9 in (23 cm).

Sedum album (white stonecrop) – a mat-forming evergreen with bead-like succulent leaves, pink stalks and branched flat-topped heads of tiny white flowers in summer; useful on walls and rocks. Any soil in a dry sunny position; 3–6 in (7.5–15 cm).

Sedum dasyphyllum – a miniature succulent with blue-green fleshy ovate leaves and small flat heads of tiny white flowers in early summer. Cultivation as above; ½ in (1 cm).

Sedum spurium 'Album' – the white form of the familiar sedum of rockeries, path edges and the tops of walls; mats of broad obovate leaves and flat heads of starry flowers in summer. Any dry sunny place; 4 in (10 cm).

Silene alba (white campion) – sticky hairy leaves and pure white flowers. Ordinary soil in sun or partial shade; 2 ft (60 cm).

Silene maritima (sea campion)* – a mound of small waxy grey-green leaves and generous numbers of white flowers in summer; 'Flore Pleno' – double; lovely over rocks and the edges of large containers. Well-drained soil in sun; 2–4 in (5–10 cm).

Silene nutans (Nottingham catchfly) – sticky leaves and one-sided clusters of night-scented white flowers with narrow, deeply cleft petals which roll back during the day; blooms from late spring to late summer. Dry calcareous stony soil in sun; 2 ft (60 cm).

Sisyrinchium idahoense (syn. *S. macounii*) 'Album' – a free-flowering miniature for rockeries and alpine troughs, with tufts of grey-green iris-like leaves and large glistening cup-shaped white flowers all summer; dislikes winter wet and benefits from shelter beneath a pane of glass. Well-drained but moisture-retentive soil in sun; 4 in (10 cm).

Sisyrinchium macounii see *S. idahoense*.

Sisyrinchium macrodenum 'May Snow' – neat clumps of grassy foliage and numerous white flowers from late spring to late summer. Well-drained peaty soil in sun; 3 in (7.5 cm).

Sisyrinchium 'Mrs Spivey' – grassy leaves and an abundance of pure white flowers. Well-drained peaty soil in sun; 9–12 in (23–30 cm).

Sisyrinchium mucronatum 'Pole Star' – miniature iris-like leaves and yellow-centred glossy white flowers all through the summer. Well-drained peaty soil in sun; 6 in (15 cm).

Smilacina racemosa (false spikenard) – rather like Solomon's seal but with fluffy heads of tiny, deliciously scented cream flowers in late spring. Rich moist soil in shade; 2–2½ ft (60–75 cm).

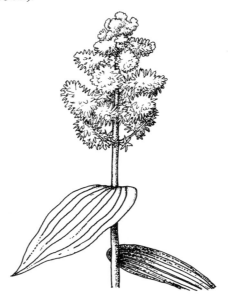

Smilacina racemosa

Smilacina stellata – (star-flowered lily-of-the-valley) – a smaller species with arching stems and terminal clusters of starry white flowers. Cultivation as above; 18 in (45 cm).

Symphyandra hofmannii (pendulous bell-flower) – a campanula relative with hairy white bells in summer. Rich well-drained soil in sun; 12–18 in (30–45 cm).

Symphytum grandiflorum (dwarf comfrey) – a vigorous ground-covering plant with rough,

dark green ovate leaves and creamy-white bell-shaped flowers in spring as the new foliage appears; 'Variegatum' – slightly smaller and less rampant, with foliage highlighted by cream markings; foolproof even under trees and shrubs. Moist soil in shade; 8–12 in (20–30 cm).

Thalictrum aquilegifolium (meadow rue) 'Album' – grey-blue pinnate leaves and panicles of fluffy white flowers in early summer. Rich moist soil in sun or light shade; 2–3 ft (60–90 cm).

Thalictrum delavayi (syn. *T. dipterocarpum*) 'Album' – a Chinese meadow rue with slightly glaucous columbine-like leaves and loose panicles of dainty little white flowers, rather like *Gypsophila*. Cultivation as above; 3 ft (90 cm).

Thalictrum dipterocarpum see *T. delavayi*.

Thymus serpyllum (creeping thyme) 'Albus'* – bright green mats and pure white flowers in summer. Well-drained soil in sun; 1 in (2.5 cm).

Tiarella collina see *T. wherryi*.

Tiarella cordifolia (foam flower) – a good evergreen ground-coverer with maple-like leaves and spikes of tiny cream flowers in early summer. Moist humus-rich soil in cool shade; 6–12 in (15–30 cm).

Tiarella trifoliata – a North American species with ivy-shaped leaves and spikes of white flowers in summer. Cultivation as above; 12–20 in (30–50 cm).

Tiarella wherryi (syn. *T. collina*) – a South American species with velvety foliage which turns russet in autumn, and brown-stalked spikes of creamy-white flowers throughout the summer. Cultivation as above; 10 in (25 cm).

Tradescantia x *andersoniana* (syn. *T. virginiana*) (trinity flower; spiderwort) – strap-shaped leaves and terminal clusters of three-petalled white flowers which are produced in succession throughout the summer; 'Innocence' – pure white; 'Osprey' – bluish-white flowers with fluffy blue centres. Well-drained but moisture-retentive soil in sun or partial shade; 18 in (45 cm).

Tradescantia virginiana see *Tradescantia* x *andersoniana*.

Tricyrtis hirta (toad lily) 'Alba' – a Japanese woodlander with erect stems and many up-right intricate greenish-white flowers in late summer and autumn. Moisture-retentive sandy soil, enriched with peat or leafmould, in shade; 18 in (45 cm).

Trientalis europaea (chickweed wintergreen) – a little woodlander with a whorl of five or six leaves from the centre of which arise flowers rather like large upward-facing pimpernels on thread-like stalks; good under conifers. Moist acid soil in dappled shade; 8 in (20 cm).

Trollius x *cultorum* (globe flower) 'Alabaster' – attractive deeply cut leaves and double cream flowers which are the nearest to white in a globe flower. Damp soil in sun or partial shade; 2 ft (60 cm). ✕

Valeriana phu 'Aurea' – bright yellow spring foliage and panicles of tubular white flowers on bare stalks in summer. Ordinary soil in sun; 2–2½ ft (60–75 cm).

Vancouveria hexandra – ivy-like foliage and sprays of tiny white flowers in early summer. Ordinary soil in shade; 12 in (30 cm).

Veratrum album (false hellebore) – a stately plant, painfully slow from seed, which eventually produces imposing clumps of magnificent pleated leaves and tall stems of small greenish-white flowers in late summer. Moist soil in partial shade; 4–6 ft (1.2–1.8 m).

Verbascum chaixii 'Album'* – a fine perennial mullein for the border, with woolly grey leaves and purple-stamened white flowers in summer. Well-drained soil in sun; 3–4 ft (90 cm–1.2 m).

Veronica gentianoides 'Alba' – a neat plant for the front of the white border; evergreen rosettes of shiny ovate to lanceolate leaves and delicate spikes of blue-budded white speedwell flowers in early summer. Moist but well-drained soil in sun or partial shade; 15–24 in (38–60 cm).

Veronica spicata 'Alba' – a reliable border plant with spikes of close-set white flowers in summer. Well-drained but moisture-retentive soil in sun; 18 in (45 cm).

Veronica virginica 'Alba' – a tall graceful North American speedwell with long tapering spikes of tiny white flowers from mid to late summer. Cultivation as above; 4 ft (1.2 m).

Vinca minor (lesser periwinkle) 'Alba' – the white-flowered form of this well-known ground-cover plant whose new spring foliage is in two-tone green; 'Alba Variegata' (syn. 'Alba Aurea-variegata') – a white-flowered form with golden variegated leaves; 'Gertrude Jekyll' – a clump-forming habit, tiny leaves and white flowers; 'Plena Alba' – the extremely rare double white form. Ordinary soil in shade; 4–5 in (10–12.5 cm) high, trailing to 18 in (45 cm) or more.

Viola – violets, violettas, violas and pansies do best in moist but well-drained humus-rich soil in sun or partial shade. They should be divided or propagated from non-flowering shoots every year or two to maintain vigour. Removing dead flowers encourages longer flowering.

Viola cornuta 'Alba'* – a species from the Pyrenees, with heart-shaped leaves and angular pure white flowers in summer; 'Alba Minor' – a small white-flowered form; 'White Perfection' – a good white form; tolerates dry shade well; 6–10 in (15–25 cm).

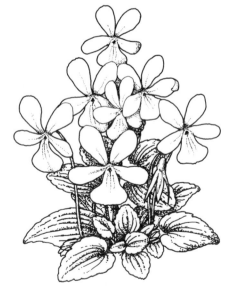

Viola cornuta *'Alba'*

Viola cucullata see *V. obliqua*.

Viola 'Janna' – a cream viola with rounded petals and a yellow eye; 4–6 in (10–15 cm).

Viola 'Lady Tennyson' – large perfectly shaped pure white violas; 6–9 in (15–23 cm).

Viola 'Little David' – a cream violetta with the scent of freesias; 4–6 in (10–15 cm).

Viola 'Milkmaid' – a prolific small bluish-white pansy; 5 in (12.5 cm).

Viola obliqua (syn. *V. cucullata*) – a North American violet with heart-shaped leaves and white to violet flowers, veined on the lower petals, in late spring and early summer; 'Alba' – large white violets with dark purple veins on smallish plants; 'Freckles' – blue-speckled bluish-white flowers; 3–6 in (7.5–15 cm).

Viola odorata (sweet violet) 'Alba' – small heavenly scented white flowers in early spring; spreads by runners; 4–6 in (10–15 cm).

Viola 'Palmer's White' – a compact old variety of pure white viola; good for bedding; 6 in (15 cm).

Viola 'Paper White' – pure white pansies with yellow eyes; 6 in (15 cm).

Viola papilionacea (Canadian violet) – white violets veined grey and blue in early summer; 3–4 in (7.5–10 cm).

Viola 'Rawson's White'* – a strongly scented little white violet which, with some protection, will bloom from late winter right through spring; 4–6 in (10–15 cm). ✕

Viola septentrionalis – large scentless white violets as the leaves emerge in spring; 6 in (15 cm).

Viola 'Swanley White'* (syn. 'Comte de Brazza') – pure white double parma violet; best under glass where it will flower profusely in winter and spring; 4–6 in (10–15 cm). ✕

Viola 'White Swan' – an old variety of bedding viola with faceless orange-eyed white blooms; 6 in (15 cm).

Viscaria alpina see *Lychnis alpina*.

Yucca filamentosa (Adam's needle) – the best species for small gardens and tubs, with stiff needle-tipped glaucous leaves and giant panicles of ivory bells in summer; 'Variegata' – a New Zealander with pink and cream variegation; flowers annually from its second or third year. Ordinary to poor well-drained soil in full sun; 2–2½ft (60–75cm), reaching well over 3ft (90cm) when flowering.

Yucca recurvifolia – a larger slower species with a woody trunk, rosettes of spiky dark green leaves and magnificent towering stems of 2–3in (5–7.5cm) creamy-white bells. Cultivation as above; up to 6ft (1.8m) tall and 7ft (2.1m) across.

Tender Perennials

Aërangis biloba – an African orchid with night-scented starry white flowers, up to 2in (5cm) across and with salmon-coloured spurs, on pendant stems 6–12in (15–30cm) long; flowers autumn to winter. Orchid compost, shade in summer, light in winter, humidity and a minimum temperature of 60°F (16°C).

Aërangis kotschyana – pendant spikes up to 18in (45cm) long in late summer bearing pure white scented flowers 1in (2.5cm) across which have spirally twisted salmon spurs. Cultivation as above.

Aërangis rhodosticta – spring to summer flowering, with orange-eyed, green-spurred white flowers. Cultivation as above.

Aërides odoratum (fox-tail orchid) – a tall-stemmed Asian orchid with strap-shaped leaves in two rows and dense pendant clusters of strongly scented white flowers in summer, variable but usually touched pink on the lip and petal tips. Cultivation as above, with a winter minimum of 55°F (13°C).

Alsobia dianthiflora see *Episcia dianthiflora*.

Angraecum distichum – a tropical West African orchid with a 10in (25cm) stem of flat ovate leaves and small starry white flowers, mostly in late summer. Orchid compost, shade in summer, light in winter, humidity and a minimum temperature of 50°F (10°C).

*Angraecum sesquipedale** – one of the most remarkable of all flowering plants, with waxy scented ivory flowers 6–7in (15–17.5cm) across which have 12in (30cm) spurs; makes a large plant with strap-shaped leaves about 12in (30cm) long; winter-flowering. Cultiva-

tion as above, maintaining a minimum temperature of 57°F (14°C).

Anguloa uniflora (syn. *A. virginalis*) – a deciduous Colombian orchid with stout pseudobulbs, bold pleated leaves and several large scented waxy tulip-like flowers in early summer; flowers variable in colour but usually white with a pink flush and pink spotted within; 'Alba' – an albino form with pure white flowers. Orchid compost; warmth, shade and humidity in summer and lighter, drier conditions and a minimum temperature of 50°F (10°C) in winter.

Anguloa virginalis see *Anguloa uniflora*.

Anthurium andraeanum (flamingo flower; painter's palette) – a tropical American aroid with handsome dark green cordate leaves and plastic-looking inflorescences which consist of a high-gloss heart-shaped spathe 4in (10cm) or more in length, and a downward-pointing straight yellowish spadix; 'Album'* – pure white; 'De Weese' – a free-flowering small white; 'Mauna Kea' – a huge cultivar with green-bordered white spathes over 8in (20cm) long; 'Manoa Mist' – a vigorous white cultivar with spathes 7in (17.5cm) long and over 5in (12.5cm) wide; 'Uniwai' – small heart-shaped white spathes with overlapping lobes, sometimes tinged pink. Does best in orchid compost with a minimum temperature of 60°F (16°C). ✕

Anthurium scherzerianum – a tougher species with leathery lanceolate leaves, waxy (but not glossy) spathes and curly spadices; 'Album'* – white spathe and bright yellow spadix. Does better as a houseplant than the above species; well-drained peat-based compost, shade, and a minimum temperature of 50°F (10°C). ✕

Brassavola nodosa (lady-of-the-night) – a Mexican orchid with fleshy cylindrical leaves about 9in (23cm) long and 3in (7.5cm) night-scented flowers which have pale green petals and sepals and a large white lip. Orchid compost and a winter minimum of 55°F (13°C).

Calanthe vestita – a deciduous tropical orchid with silvery 6in (15cm) pseudobulbs, ribbed lanceolate leaves and 2–3ft (60–90cm) spikes bearing twenty or more variable flowers (usually white with a pink lip) from autumn to winter after the leaves have died off; 'Nivalis' –

a Javanese variety with pure white flowers. Repot annually as new growth begins in spring, using a fibrous compost enriched with peat, leafmould, silver sand and well-rotted manure; needs sun all year round; keep dry and at a minimum temperature of 55°F (13°C) when dormant; provide ample warmth, moisture and humidity when in growth.

Campanula isophylla – a dwarf Italian species with toothed heart-shaped leaves and 1in (2.5cm) starry flowers in summer; 'Alba' – the original white form which cannot be grown from seed and has to be propagated from stock plants; 'Kristal White'* – a vigorous new white variety which comes true from seed; 'Stella White' (F1 hybrid) – another floriferous new variety which produces superb plants from seed. Peat-based compost in sun or partial shade; 6in (15cm) high and 12in (30cm) or more across.

Catharanthus roseus (syn. *Vinca rosea*) (Madagascan periwinkle) – small glossy ovate leaves and a succession of flat five-petalled flowers from spring to autumn; 'Albus' – pure white with a pale yellow eye. There are also white cultivars with bright pink eyes. Peat-based compost in sun with a winter minimum of 45°F (7°C); 12in (30cm).

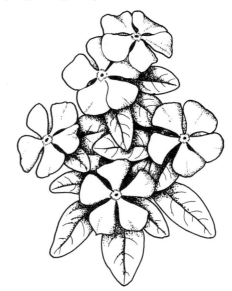

Catharanthus roseus

Cattleya 'Bow Bells'* – an old but still outstanding hybrid derived from albino forms of *C. trianae* and *C. mossiae*; huge pure white flowers with yellow-marked lips. Orchid compost with shade, ample warmth and humidity in summer, with a cooler, drier, lighter winter rest at a minimum temperature of 50°F (10°C). ✕

Cattleya skinneri 'Alba'* – an elegant Guatemalan orchid which bears two leaves atop a 9 in (23 cm) pseudobulb and long-stemmed spikes of five to nine ivory flowers, marked yellow on the lip and about 3 in (7.5 cm) across; spring-flowering. Cultivation as above. ✕

*Coelogyne cristata** – an easy evergreen Himalayan orchid with bright green egg-shaped pseudobulbs, pairs of leathery lanceolate leaves, and long-lasting pendant spikes of two to seven wavy-petalled flowers, ice white with egg-yolk yellow striations on the lip and up to 4 in (10 cm) across; winter-flowering; 'Alba' – pure white throughout; 'Lemoniana' – lemon-yellow markings on lip. Orchid compost (but repot infrequently); cool dry bright winter rest at a minimum 50°F (10°C) (the pseudobulbs normally shrivel but this doesn't matter); warmth, moisture, humidity and shade in summer.

*Coelogyne ochracea** – a neat easy little Indian orchid with compact growths 8–10 in (20–25 cm) high and numerous small white flowers, marked yellow and orange on the lip, in spring; 'Alba' – pale yellow markings. Cultivation as above.

Cymbidium hybrids – narrow arching leaves and tall spikes of large waxy flowers which last two months or more; available as full-size plants about 3 ft (90 cm) tall with flowers up to 4 in (10 cm) across, or as miniatures at half the size. Full size: 'Pearl Balkis' – white to pink with crimson on the lip; 'Pharoah' – white with pale cream lip markings; 'Sleeping Giant'* – pure white. Miniatures: 'Showgirl'* – white to pink. Plants should be chosen in flower, as named white to pink cultivars may vary in colour from plant to plant. Orchid compost with warmth, humidity and shade in summer followed by a cooler, drier, lighter winter rest at a minimum 45°F (7°C). ✕

Cymbidium pumilum 'Album' – not exactly white, but a very pale form of a dwarf Japanese species which bears erect spikes of small brownish-red flowers; an important 'stud' plant for breeders of miniature cymbidiums. Cultivation as above.

Dendrobium kingianum 'Album' – an evergreen Australian orchid with clusters of leathery leaves atop club-shaped pseudobulbs up to 20 in (50 cm) long and apical spikes of small, half-opening but very pretty pure white flowers in spring; cultivation as for cymbidiums though tolerant of lighter, drier conditions.

Dendrobium nobile 'Album'* (syn. *D. nobile* var. *virginalis*) – the albino form of the most popular dendrobium in cultivation; Asian in origin and evergreen, with cane-like pseudobulbs up to 18 in (45 cm) tall and clusters of two to three flowers, each 2–3 in (5–7.5 cm) across, at the nodes in late winter; scented and varying in odour according to the time of day; orchid compost with plenty of warmth, moisture, humidity and shade during the growing season, followed by a pronounced winter rest in light dry conditions at a minimum 55°F (13°C).

Dendrobium wardianum 'Album' – the albino form of a deciduous Burmese species with 12–24 in (30–60 cm) pseudobulbs, along which are borne clusters of long-lasting 3–4 in (7.5–10 cm) flowers, often numbering over thirty per pseudobulb, in spring; cultivation as above.

Epidendrum fragrans – a tropical American orchid with 3–4 in (7.5–10 cm) pseudobulbs, each bearing a single leaf and, from late summer to early winter, a spike of two to six upside-down flowers with scented ivory to greenish-white petals and sepals and a pink-pencilled lip; orchid compost and plenty of warmth, moisture, shade and humidity in summer, followed by cooler, drier, lighter conditions in winter at a minimum 55°F (13°C).

Epiphyllum x *ackermannii* (orchid cactus) – free-flowering hybrids derived mainly from *E. cooperi*, a shy-flowering tropical rain forest cactus with long, flattened fleshy stems and large creamy-white scented flowers, in appearance midway between a clematis and a water lily; 'Belgica' – creamy-white; 'Chantal' – white; 'Duke of York' – large cream; 'William Rohbock' – white. Peat-based compost with shade, moisture and warmth in summer, followed by a cooler, drier, brighter rest in winter at a minimum 45°F (7°C).

*Episcia dianthiflora** (syn. *Alsobia dianthiflora*) – rosettes of small dark green velvety leaves and large crystalline white flow-

Episcia dianthiflora

ers with deeply fringed petals, resembling snowflakes; spreads by runners. Peat-based compost with extra sand; needs warmth, shade and humidity with a winter minimum of 60°F (16°C).

Hedychium coronarium (white ginger; Indian garland flower; butterfly lily) – a deciduous rhizomatous species with cane-like stems of lanceolate leaves and terminal clusters of pure white, heavily scented orchid-like flowers in summer. John Innes No. 3 with ample warmth and water in sun or partial shade during growth, with a dryish rest in winter at a minimum 45°F (7°C).

Hoya bella (wax flower) – an Indian species with arching green twiggy stems, small fleshy ovate leaves and pendant umbels of fragrant pink-eyed white flowers, waxy in texture and shaped like blunt five-pointed stars; best in a hanging basket or wall pot where you can look up at the flowers; peat-based compost with warmth, humidity and light shade in summer and a dryish sunny rest in winter at a minimum 55°F (13°C).

Impatiens sultanii (busy lizzie) – an evergreen perennial from Zanzibar, now superseded by hybrids which are usually grown from seed and treated as hardy annuals, though they can be overwintered or propagated from cuttings; finely toothed elliptic leaves and flat six-petalled flowers with curved spurs; 'Futura White' – numerous large white flowers on compact semi-pendant plants; 'Super Elfin White'* – quick and easy from seed, floriferous and large-flowered. Peat-based compost in sun or partial shade with a minimum temperature of 55°F (13°C).

Laelia anceps 'Alba'* – the albino form of an evergreen Mexican orchid which produces elegant yellow-throated ivory flowers, two to five in number and about 4 in (10 cm) across, on erect wiry 2 ft (60 cm) stems during winter and early spring; orchid compost and ample warmth, moisture, humidity and shade in summer, followed by a cool, sunny, dry rest in winter and a minimum temperature of 50°F (10°C); repot only when the compost has broken down.

Lycaste skinneri (syn. *L. virginalis*) 'Alba'* – the pure white form of a Central American orchid which has 8 in (20 cm) pseudobulbs, pleated oblong-lanceolate deciduous leaves and fragrant waxy flowers, roughly triangular in outline and some 6 in (15 cm) across; winter-flowering during dormancy. Orchid compost and warm, shady, humid conditions in summer, followed by a cool dry sunny rest at a minimum of 50°F (10°C).

Lycaste virginalis see *Lycaste skinneri*.

Masdavallia tovarensis – a curious little Venezuelan orchid with clumps of fleshy leaves and 6 in (15 cm) spikes of three to four squarish long-tailed matt white blooms in winter. Old spikes may flower again the next year so do not remove. Orchid compost with warmth, shade and humidity all year with a winter minimum of 50°F (10°C).

Nymphaea lotus (Egyptian lotus) – a tropical night-flowering Egyptian water lily with distinctly toothed leaves and narrow-petalled matt white flowers – 6–8 in (15–20 cm) wide, which are held well above the water; needs a pool with at least 3 in (7.5 cm) of rich mud and a 12 in (30 cm) depth of water, maintaining a minimum summer temperature of 70°F (21°C); gradually dry out and store in moist sand at 50°F (10°C) during the winter, or continue heating the pool to this temperature. ✕

Odontoglossum citrosmum 'Album' – the white form of a lemon-scented Mexican orchid which has pairs of broad leathery leaves on each 3–5 in (7.5–12.5 cm) pseudobulb and in summer produces long pendant spikes of up to thirty white flowers, marked yellow on the lip and over 2 in (5 cm) across. Orchid compost with airy, moist conditions in summer followed by a pronounced dry cool bright rest in winter (the pseudobulbs will shrivel but subsequently recover when growth and watering recommence). Maintain a minimum temperature of 50°F (10°C).

*Odontoglossum crispum** – pairs of evergreen lanceolate leaves from the apex of ovate 4 in (10 cm) pseudobulbs and arching basal spikes, 2 ft (60 cm) long, bearing variable broad-petalled white flowers, 4 in (10 cm) across, which are usually spotted with reddish-brown and sometimes pink-flushed; var. *xanthotes* – the albino form with yellow-spotted white flowers (even whiter is the occasional unspotted clone of the species, though such plants are rare owing to the fact that they were once discarded as inferior to those with the familiar reddish markings); mainly spring-flowering, dislikes excessive warmth and 'mugginess'. Orchid compost with shade, humidity and good ventilation in summer and a cool, bright, drier (but not dry at the roots) winter rest at a minimum temperature of 50°F (10°C). ✕

Oncidium ornithorhynchum 'Album' – the white-flowered form of a small leafy Central American orchid which produces generous numbers of branched wiry spikes in autumn and winter, laden with crinkled little flowers which have a surprisingly powerful scent. Orchid compost with shade, humidity and warmth in summer, followed by a winter rest with maximum light, just sufficient moisture to prevent shrivelling and a minimum temperature of 50°F (10°C).

Osteospermum tauranga 'Whirligig'* – an eyecatching African daisy with steely blue-centred, blue-backed brilliant white flowers which have each petal neatly pinched in the middle; very easy from cuttings; 2 ft (60 cm). John Innes No. 1 in sun with a winter minimum of 45°F (7°C).

Paphiopedilum (slipper orchid) 'F. C. Puddle' – a large white hybrid which was a milestone in orchid breeding and parent of many later white-flowered 'paphs'. Orchid compost with warmth, shade and humidity all year, with a winter minimum of 60°F (16°C). ✕

Paphiopedilum 'Maudiae' – another important early hybrid with mottled leaves and long-stemmed green and white striped flowers 4 in (10 cm) across which last two months. Cultivation as above. ✕

Paphiopedilum fairrieanum 'Album' – the albino form of a delightful slipper orchid from the Himalayas which has tufts of light green leaves and long-lasting quaint flowers in autumn whose petals are crimped at the margin and neatly curled. Instead of green and white with maroon veins, the albino is white with yellow striping. Orchid compost with warmth, humidity and shade in summer, followed by cooler, rather drier conditions in winter at a minimum temperature of 50°F (10°C). ✕

Paphiopedilum niveum 'Album' – the unspotted form of a white-flowered slipper orchid which has handsome marbled grey-green leaves with purple undersides and short-stalked, fat little pouched flowers that last for months. Orchid compost with heavy shade, ample warmth and humidity in summer and a minimum temperature of 60°F (16°C) in winter.

Pelargonium x *domesticum* (regal pelargonium) – an evergreen sub-shrub with toothed ovate to palmate leaves and clusters of large wavy flowers, each measuring 2 in (5 cm) or more across, in summer; 'Nomad' – white with crimson markings on the upper petals; 'White Glory'* – pure white. John Innes No. 2 with full sun and moderate watering in summer, keeping almost dry and at a minimum 45°F (7°C) in winter, cutting back and repotting in spring; 2 ft (60 cm) or more.

Pelargonium x *hortorum* (geranium; zonal pelargonium) – familiar and reliable pot plants with light green rounded leaves which have a characteristic bronze band (usually absent in white-flowered cultivars) and dense umbels of dazzling flowers from spring to autumn, and even through the winter; 'Avalon' – dwarf white; 'White Century'* – single pure white; 'Hermione' – double white; 'Snow Queen' – double white with a pink flushed centre; 'White Orbit' – compact single with faintly zoned leaves. Cultivation as above; 9 in (23 cm) upwards.

Pelargonium peltatum (ivy-leaved geranium) – a pendant species with brittle palmate leaves and clusters of vivid flowers from early spring to late autumn; 'L'Elégante'* – a distinctive cultivar which dates back to Victorian times, with white-margined leaves which flush lilac in autumn and large lilac-white flowers; 'Snowdrift' – white. Cultivation as above; 12 in (30 cm) or more.

Peristeria elata (dove orchid; Holy Ghost flower) – a Central American orchid with very large pseudobulbs, ribbed lanceolate leaves up to 3 ft (90 cm) tall and equally long erect spikes of scented waxy white flowers whose complex lip is purple-speckled and resembles a hovering dove; summer-flowering. Orchid compost with warmth, humidity and shade in the growing season and a cooler, brighter, dryish winter rest at a minimum 60°F (16°C).

Phalaenopsis amabilis (moth orchid)* – a tropical Asian species, unscented, but one of the loveliest flowers on earth; thick oblong-lanceolate semi-pendant leaves and long arching spikes of a dozen or more 4 in (10 cm) long-lasting pure white flowers, finely marked red and yellow inside the lip; generally autumn and winter flowering. Orchid compost with shade, ample warmth, moisture and humidity in summer, and rather less of everything in winter, at a minimum temperature of 60°F (16°C).

Phalaenopsis hybrids – easier to grow as house or conservatory plants than the species, with a great range of superb white-flowered varieties, including 'Aubrac' and 'Henrietta Lecoufle'*. Cultivation as above.

Phalaenopsis stuartiana – mottled foliage and long branched spikes of white flowers which are gold-flushed and brown-speckled on the ornate lip and lower halves of the lateral petals; flowers at various times. Cultivation as above.

Saintpaulia ionantha (African violet) – familiar dark green hairy roundish leaves surmounted by a posy of long-lasting yellow-eyed 'violets'; 'White Fairy Tale'* – an F1 hybrid with single pure white flowers, a refreshing change from the messy bicolours, frills and attempts at variegation which are currently blighting these delightful pot plants. Peat-based compost in light shade with a winter minimum of 55°F (13°C).

Selenicereus grandiflorus (torch thistle; night-flowering cereus)* – a West Indian cactus with spiny rope-like stems and huge creamy-white multi-petalled flowers which are heavily scented and open at dusk, only to die at dawn; summer-flowering on mature plants. John Innes No. 2 with additional coarse grit with ample warmth and moisture in the summer but a cool dryish rest at a minimum 45°F (7°C) in winter.

Selenicereus grandiflorus

Spathiphyllum 'Mauna Loa'* – a hybrid peace lily which is larger all round than *S. wallisii* and flowers mainly in spring, with the occasional inflorescence throughout the year. Cultivation as below, winter minimum 55°F (13°C).

Spathiphyllum wallisii (peace lily)* – a Colombian aroid with shiny rich green lanceolate leaves and showy inflorescences in summer, consisting of an erect white ovate spathe and a spiky cream spadix. Peat-based compost with ample warmth, shade and moisture during the summer and a minimum temperature of 50°F (10°C) in winter.

Streptocarpus x *hybridus* (Cape primrose) – a reliable summer flowering pot plant with strap-shaped suedette leaves and numerous large flared tubular flowers; 'Albatross'* – pure white flowers, lightly marked lemon yellow in the throat; 'White Concorde' – flowers within four months from seed; 'Lipstick White' – pure white flowers with finely frilled margins and purple markings in the throat. Peat-based compost in partial shade with a winter minimum of 50°F (10°C).

Verbena teucrioides – a Brazilian species, parent of bedding verbenas, with hairy heads of large blush-white or yellowish-white flowers which release a lemon scent in the evening. John Innes No. 2 in partial shade, with a winter minimum of 40°F (5°C); 12 in (30 cm).

Victoria amazonica (royal water lily) – one of the most spectacular white-flowered plants in the world, worth converting your heated swimming pool into a tropical lily pond to accommodate; leaves the size of table-tops, with upturned rims and prickly ribbed undersides, on stalks over 10 ft (3 m) long and prickly-budded multi-petalled flowers, some 12 in (30 cm) across, which are flushed pink in the centre. The seed (if you can get it) must be kept constantly wet and germinated in pots of rich soil, submerged in a tub or tank at 85°F (30°C) during the winter. The seedling should be potted on as growth accelerates until ready for 'launching' into the pool. Needs rich mud, full sun and high temperatures all year round for perennial growth, otherwise is best treated as an annual.

Vinca rosea see *Catharanthus roseus.*

Hardy and Near Hardy Shrubs

Abutilon vitifolium 'Album' and 'Tennant's White'* – beautiful near-hardy shrubs with felted palmate leaves and flat white flowers; needs a warm sheltered position. Ordinary well-drained soil in sun; 8 ft (2.4 m).

Amelanchier canadensis (shadbush; June berry; snowy mespilus) – a suckering shrub with downy toothed leaves, masses of star-shaped white flowers in spring before the new foliage, followed by edible black berries and good autumn colour. Moisture-retentive soil in sun or partial shade; 10 ft (3 m).

Amelanchier lamarckii 'Ballerina' – rich coppery new leaves with white flowers in spring, followed by edible red berries and purplish-bronze autumn colour; cultivation as above; 6–8 ft (1.8–2.4 m).

Berberis x *stenophylla* 'Cream Showers' – the nearest yet to a white-flowered berberis, with arching branches of clustered cream bells in spring; evergreen, hardy and tolerant of poor thin soils in sun or partial shade; 7–8 ft (1.2–2.4 m).

Buddleia davidii (butterfly bush) – a vigorous deciduous shrub with lanceolate leaves and fragrant conical heads of tiny tubular flowers which attract hosts of butterflies in summer; 'Peace', 'White Cloud' and 'White Profusion' are some of the growing numbers of white-flowered cultivars. Cut back hard in spring for strong leafy branches and large flower heads; ordinary soil in sun; 9 ft (2.7 m) or more.

Buddleia fallowiana 'Alba' – a deciduous shrub with white woolly branches, silvery grey lanceolate leaves and long clusters of creamy-white, orange-eyed scented flowers in summer; hardier than the usual lilac form but best in a sunny sheltered position; 5–10 ft (1.5–3 m).

Calluna vulgaris (ling; heather) – a small wiry shrub with scale-like leaves and dense colourful spikes of flowers over a long period during late summer, autumn and winter; 'Alba Plena' – double white; 'Alba Pumila' – a white-flowered miniature; 'Alba Rigida' and 'White Lawn'* – prostrate whites; 'Hammondii' – fine dark foliage and dense white flowers; 'Kinlochruel'* – pure white double flowers and a compact habit. Lime-free soil in sun; 12–18 in (30–45 cm), with the exception of prostrate forms which reach about 2 in (5 cm). ✕

Camellia – though generally hardier than supposed, camellias flower so early that they need protection from cold winds, severe frosts and early morning sun, all of which can damage buds and blooms. In addition, they must have lime-free humus-rich soil with a cool root run and good drainage. All white-flowered camellias are more sensitive to frost than the reds.

Camellia 'Cornish Snow' – a hybrid between *C. cuspidata* and *C. saluensis*, with glossy elliptic leaves and abundant small flowers clustering along the branchlets from late winter to spring; 8–10 ft (2.4–3 m).

Camellia japonica – an upright shrub with glossy, leathery ovate leaves and 3–5 in (7.5–12.5 cm) flowers at the ends of the branches in late winter. There are innumerable white varieties including: 'Alba Plena' – large double white flowers whose petals are neatly overlapping (known in the camellia world as a formal double); 'Alba Simplex' – large flat single white flowers with conspicuous yellow stamens; 'Devonia' – small semi-double with some resistance to browning; 'Haku-rakuten' – large semi-double (peony-flowered); 'Matho-

tiana Alba'* – large double white (formal shape), less hardy than most; 'Nobilissima' – an early and rather sensitive variety with peony-shaped white flowers which shade to yellow; 'Purity' (syn. 'Shiragiku') – formal double white, shading to cream in the centre; 'Silver Anniversary' – huge double peony-flowered; 'White Giant' – huge double; 'White Nun' – large white semi-double; 6–10 ft (1.8–3 m). ✕

Camellia x *williamsii* – a hybrid between *C. saluensis* and *C. japonica* which flowers when very young and sheds faded flowers (unlike most camellias which retain brown and dying blooms); 'China Clay' – an open shrub with semi-double white flowers; 'E. T. R. Carylon'* – a late-flowering white with loosely double flowers; 'Francis Hanger' – an upright shrub with single white flowers. Best in sheltered woodland conditions or against a wall; 6–8 ft (1.8–2.4 m). ✕

Carpenteria californica – a reasonably hardy evergreen shrub, given a sheltered site; shiny lanceolate leaves and scented white anemone-like flowers with yellow stamens in summer; 'Bodnant'* – larger flowers. 10 ft (3 m).

Carpenteria californica

Cassiope lycopodioides – a prostrate, hardy evergreen shrublet with wiry dark green scaly branches and exquisite white bells in spring. Cool acid soil in a lightly shaded but open position; 2–3 in (5–7.5 cm).

Cassiope tetragona – similar to the above but upright in habit; 12 in (30 cm).

Cercis siliquastrum (Judas tree) 'Alba' – a hardy deciduous small tree or spreading shrub with clusters of white pea flowers on naked twigs in late spring, followed by almost round leaves and long reddish seed pods. Resents disturbance and needs a warm sheltered site to protect the flowers from sharp frosts; ordinary soil in sun; 15 ft (4.5 m) or more.

Chaenomeles speciosa (Japanese quince) 'Nivalis'* – a hardy deciduous shrub whose pure white flowers show to advantage when it is trained against a red brick wall; blooms from winter to spring before the new leaves appear. Ordinary soil in sun; 6 ft (1.8 m) or more.

Chamaepericlymenum canadensis see *Cornus canadensis*.

Choisya ternata (Mexican orange blossom)* – a reasonably hardy evergreen shrub with aromatic, shiny trifoliate leaves and clusters of scented white flowers in late spring and early summer; 'Sundance' – yellow foliage. Well-drained soil in a sheltered position in sun or partial shade; 5–6 ft (1.5–1.8 m).

Cistus – sun roses are a doubtful proposition for areas subject to hard winters and cold winds. They need full sun and do best on poor well-drained to dry soils. Planting should be done with the minimum of disturbance.

Cistus x *aguilari* – shiny evergreen leaves with wavy margins and large pure white flowers in summer; not hardy in cold areas; 4–5 ft (1.2–1.5 m).

Cistus x *corbariensis* – one of the hardiest sun roses; spreading habit, dull ovate leaves and small, yellow-centred flowers which are crimson-flushed in bud; 3–4 ft (90 cm–1.2 m).

Cistus x *cyprius* – sticky foliage and large flowers with crimson blotches at the base of each petal; 'Elma' – larger flowers and more compact growth; 6–8 ft (1.8–2.4 m).

Cistus ladanifer – a reasonably hardy upright shrub with narrow sticky leaves and large ruffled white flowers which have a boss of golden stamens and a five-pointed crimson star in the centre; 5–6 ft (1.5–1.8 m) or more.

Cistus x *lusitanicus* – narrow dark green lanceolate leaves and white flowers with a maroon blotch at the base of each petal; 'Decumbens' – a variety with a spreading habit, reaching 4 ft (1.2 m) across; 12–24 in (30–60 cm).

Cistus palhinhaii – a Portuguese sun rose with sticky foliage and large crinkled white flowers, 4 in (10 cm) across, in early summer; 3 ft (90 cm).

*Cistus populifolius** – shiny, sticky leaves and 2 in (5 cm) white flowers, stained yellow in the centre; var. *lasiocalyx* has larger flowers; up to 6 ft (1.8 m).

Cistus salvifolius – reasonably hardy in mild areas and good ground-cover on dry poor soils; greyish, sage-like leaves and yellow-centred white flowers in early summer; 12–24 in (30–60 cm).

Clethra alnifolia (sweet pepper bush) – a deciduous upright shrub with serrated oblong leaves and terminal spikes of fragrant creamy-white flowers in summer; 'Paniculata' – narrower leaves and larger flowers. Moisture-retentive lime-free soil in sun or light shade; 6 ft (1.8 m).

Clethra delavayi – a Chinese species with deeply veined lanceolate leaves and horizontal one-sided spikes, 6–8 in (15–20 cm) long, of small creamy-white flowers which have pink calyces. Moist but well-drained lime-free soil in a sheltered position; about 15 ft (4.5 m).

Colletia armata (anchor plant) – a truly ferocious shrub from Chile, consisting of huge green triangular spines and, in early autumn, masses of tiny scented white bells which are alive with bees. Needs a warm sheltered site; well-drained soil in sun; 10 ft (3 m).

*Convolvulus cneorum** – a delightful small evergreen shrub with silky silver foliage and white bindweed flowers which are pink in bud and open throughout the summer. Not reliably hardy; well-drained soil in full sun; 2–3 ft (60–90 cm).

Cornus canadensis (syn. *Chamaepericlymenum canadensis*) (creeping dogwood) – more like a herbaceous perennial than a shrub, with a ground-covering habit; insignificant flowers surrounded by showy white bracts in summer, followed by red berries. Lime-free soil, enriched with peat, in light shade; 6 in (15 cm).

*Cornus kousa** – an oriental dogwood with wavy-margined, leaves and purplish-green flowers in early summer, each surrounded by four white bracts and followed by raspberry-like fruits and good autumn colour; var. *chinensis* – the Chinese form which is taller, more upright and open, and has larger bracts. Well-drained soil in sun; 8–10 ft (2.4–3 m).

Cornus nuttallii – an American dogwood with ovate leaves and numerous flowers in late spring, each consisting of a central bobble of inconspicuous flowers surrounded by four or more white bracts up to 3 in (7.5 cm) long which flush pink as they age. Cultivation as above; up to 20 ft (6 m).

Cornus nuttallii

Cotoneaster – the majority of cotoneasters have white flowers but are mainly grown for their (mostly) bright red berries. The only ones I have included are unusual varieties with yellowish fruits. They will grow in almost any soil in sun or partial shade.

Cotoneaster frigidus – a quick-growing semi-evergreen Himalayan shrub or small tree with oblong leaves and clusters of tiny white flowers in early summer, followed by heavy long-lasting bunches of berries; 'Fructu-luteo' – creamy yellow berries. 12–15 ft (3.6–4.5 m).

Cotoneaster x *rothschildianus* – a semi-evergreen spreading shrub with light green lanceolate leaves, flat clusters of tiny white flowers in early summer, followed by dense masses of rich cream berries; up to 15 ft (4.5 m) high and 10 ft (3 m) across.

Cotoneaster x *watereri* – a semi-evergreen shrub or small tree with an arching habit, dark green lanceolate leaves, clusters of tiny white flowers in early summer, and large bunches of long-lasting berries; 'Pink Champagne' – yellow pink-tinged berries; 12–15 ft (3.6–4.5 m).

Crataegus laevigata (syn. *C. oxycantha*) (Midland hawthorn) – very similar to common hawthorn but with rounder, more shallowly lobed leaves; 'Plena' – double flowers. Ordinary soil in sun; up to 20 ft (6 m).

Crataegus x *lavallei* – a hybrid thorn with virtually no thorns, a densely branched habit, shiny ovate leaves and erect clusters of white flowers in early summer, followed by long-lasting reddish-orange haws. Cultivation as above; up to 20 ft (6 m).

Crataegus monogyna (hawthorn; quickthorn; may) – a thorny, densely branched shrub or small tree with lobed and toothed leaves, clusters of heavily scented white flowers in late spring, followed by dark red haws; 'Biflora' (syn. 'Praecox') (Glastonbury thorn) – comes into leaf very early and flowers in winter during mild weather; 'Compacta' – a compact thornless cultivar; 'Stricta' – erect habit. Cultivation as above; 25 ft (7.5 m) or more.

Crateagus oxycantha (*oxycanthoides*) see *C. laevigata*.

Cytisus albus see *Cytisus multiflorus*.

Cytisus multiflorus (syn. *C. albus*) (white broom)* – an erect grey-green shrub with arching branches of white flowers in spring; resents disturbance and tends to be short-lived on alkaline soil. Ordinary to poor soil in sun; 6 ft (1.8 m).

Cytisus x *praecox* (Warminster broom) 'Albus' – a dense arching grey-green shrub which is completely covered in white flowers in spring. Cultivation as above; 5–6 ft (1.5–1.8 m).

Cytisus purpureus 'Albus' – the white-flowered form of purple broom, a small shrub with dark green trifoliate leaves in late spring. Cultivation as above; 12–24 in (30–60 cm) high and about 4 ft (1.2 m) across.

Cytisus scoparius 'Cornish Cream' – the nearest to white in the common broom, with bicoloured cream and white flowers in late spring. Cultivation as above; up to 8 ft (2.4 m).

Daboecia cantabrica (St Dabeoc's heath) – a heather-like shrub with dark green foliage and large globular bell flowers from late spring to winter; 'Alba' – white-flowered; 'Alba Globosa' – fatter white flowers. Peaty lime-free soil in sun or partial shade; 3 ft (90 cm).

Daphne blagayana – a small species which forms mats of spatulate leaves and bears clusters of intensely fragrant creamy-white flowers in spring. Well-drained neutral to alkaline soil in sun or partial shade; 6 in (15 cm) high and up to 6 ft (1.8 m) across.

Daphne cneorum (garland flower) 'Alba' – a prostrate evergreen plant with dark green oblong leaves and clusters of scented white flowers in early summer; less vigorous than the species and its other varieties. Cultivation as above; 6 in (15 cm) high and 2–3 ft (60–90 cm) across.

Daphne mezereum (mezereon) 'Alba' – a deciduous shrub which produces very fragrant flowers close to the bare twigs in late winter and early spring, followed by poisonous yellow berries. Cultivation as above; 5–6 ft (1.5–1.8 m), though usually about half this height.

Daphne odora 'Alba'* – an evergreen Oriental species of doubtful hardiness which bears ovate shiny leaves and clusters of white flowers, smelling of lemons and roses, in late winter. Needs a sheltered position and winter protection; excellent in pots for the cool conservatory. Cultivation as above; 5 ft (1.5 m).

Davidia involucrata* (handkerchief tree; dove tree; ghost tree) – a hardy deciduous Chinese tree of stupendous beauty when flowering in late spring, with knots of greenish flowers from which are suspended huge white bracts, 7 in (17.5 cm) long, that wave in the breeze; the perfect centrepiece for a sizeable lawn. Moisture-retentive soil in sun or partial shade; up to 25 ft (7.5 m) tall and 18 ft (5.5 m) wide.

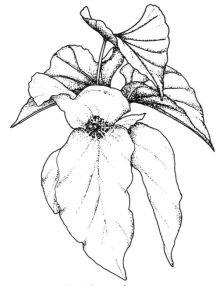

Davidia involucrata

Deutzia gracilis – a hardy deciduous Japanese species with pointed lanceolate leaves and generous numbers of starry white flowers in early summer. Well-drained soil in sun or partial shade; 3–4 ft (90 cm–1.2 m).

Deutzia x *magnifica* – an upright hybrid with grey-green foliage and 3 in (7.5 cm) panicles of double white flowers in early summer. Cultivation as above; 8 ft (2.4 m).

Deutzia scabra 'Candidissima' – the double white form of an Oriental species which has the added attraction of peeling bark. Cultivation as above; 6–10 ft (1.8–3 m).

Deutzia setchuenensis var. *corymbiflora* – a Chinese species with peeling bark, toothed leaves which are downy underneath, and 4 in (10 cm) clusters of starry white flowers in summer. Cultivation as above; 6 ft (1.8 m).

Erica – heaths or heathers are easy evergreen shrubs with year-round interest. They prefer an open sunny position in damp peaty soil. Some dislike lime. Individual requirements should be checked before purchase.

Erica arborea (tree heath) – a reasonably hardy Mediterranean species with racemes of fragrant off-white flowers in spring; 'Alpina' – a hardier cultivar with a bushy habit and bright green foliage; 'Estrella Gold' – golden foliage. 4–5 ft (1.2–1.5 m).

Erica carnea (heath; heather) – a dense evergreen shrub with needle-like leaves and terminal racemes of tiny bell-shaped flowers from autumn to spring; 'Alba' – white; 'Springwood White'* – a vigorous white-flowered variety, suitable for ground-cover; 10–12 in (25–30 cm); lime-tolerant.

Erica ciliaris (Dorset heath) – light green foliage and bell-shaped flowers from summer to winter; 'Alba' – white; 12–18 in (30–45 cm). 'Stoborough' – a tall white variety, reaching 2 ft (60 cm).

Erica cinerea (bell heather) – dark green leaves and rounded bell flowers from early summer to autumn; 'Alba Minor' – a compact white-flowered variety, reaching only 6 in (15 cm); 'Hookstone White'* – a vigorous white variety; 18 in (45 cm); 'White Dale'* – brilliant white flowers; 12 in (30 cm); tolerates drier conditions than most ericas.

Erica x *darleyensis* – a lime-tolerant hybrid heather with creamy-white to pink new growths and flowers from winter to late spring; 'Ada S. Collins'* – dark foliage and white flowers, 8 in (20 cm); 'Silberschmelze' (syn. 'Alba'; 'Molten Silver')* – silvery-white flowers over a long period, 18 in (45 cm).

Erica erigena (syn. *E. mediterranea*) – a reasonably hardy large heath which bears long branched racemes of flowers from winter to summer; 'Alba' – white; 'Alba Compacta' – a smaller denser white form; 'W. T. Rackliff'* – mounds of rich green foliage and creamy-white flowers; 2 ft (60 cm).

Erica mediterranea see *Erica erigena*.

Erica tetralix (cross-leaved heath) – grey-green hairy leaves and clusters of bells in summer; 'Alba' – white; 'Alba Mollis'* – silver-grey leaves and white flowers; 'Alba Praecox' – an early white form; prefers damp to wet conditions; 8–12 in (20–30 cm).

Erica vagans (Cornish heath) – a large heath which flowers from summer to winter and tolerates slightly alkaline soil; 'Cream'* – dense bushes of creamy-white flowers, 20 in (50 cm); 'Lyonesse'* – bright green leaves and brown-anthered white flowers, 18 in (45 cm); 'Valerie Proudley' – a slow-growing cultivar with golden foliage and white flowers, 6–10 in (15–25 cm).

Erica x *veitchii* – a hybrid tree heath, hardy only in mild areas; 'Exeter' – scented white flowers in winter and early spring; 8 ft (2.4 m).

Escallonia 'Iveyi' – a hybrid with glossy evergreen ovate leaves and 5–6 in (12.5–15 cm) panicles of white tubular bells in summer; not hardy in cold areas. Well-drained neutral to alkaline soil in a sunny sheltered positiion; does well in coastal situations; 10 ft (3 m).

Eucryphia x *nymansensis* 'Nymansay'* – a supremely beautiful evergreen tree which is quick-growing, reasonably hardy in mild areas, and almost completely covered in late summer with cream flowers which are 2½ in (6.5 cm) across and filled with stamens (and bees); needs winter protection when young. Cool neutral to acid soil in a sheltered sunny or partially shaded position; 15 ft (4.5 m) tall and 6–8 ft (1.8–2.4 m) wide.

Exochorda giraldii (pearl bush) – a vigorous shrub with arching branches laden with large, short-lived white flowers in spring. Ordinary soil in sun or partial shade; 15 ft (4.5 m).

Exochorda x *macrantha* (pearl bush) 'The Bride' – a hardy deciduous shrub with a compact weeping habit and masses of large white flowers in late spring. Cultivation as above; 10 ft (3 m).

Fothergilla gardenii – a hardy deciduous American shrub, related to witch hazel; scented off-white flowers in short bottle-brush clusters in spring; hazel-like leaves and brilliant autumn colour; very effective against a dark background such as conifers. Moist humus-rich lime-free soil; 3 ft (90 cm).

Fothergilla major – similar to the above but larger all round and with an upright pyramidal habit. Cultivation as above; 9 ft (2.7 m).

Fothergilla monticola (Alabama fothergilla) – slightly larger flowers and a more spreading habit than *F. major*, otherwise similar in appearance and cultivation; 6 ft (1.8 m).

Franklinia alatamaha – a shapely deciduous shrub or tree with conspicuous yellow-centred fragrant white flowers, 3 in (7.5 cm) across and resembling camellias, from late summer to autumn when the leaves turn bright red. Needs protection in severe winters. Ordinary soil in sun; 25 ft (7.5 m).

Fuchsia magellanica 'Alba'* – a reasonably hardy deciduous shrub with slender branches and small, graceful, blush-white flowers; well-drained soil in sun or partial shade; may be cut back by frost but usually sprouts again from the base; can reach 10 ft (3 m) but in colder areas may be half this or less.

Gaultheria procumbens (wintergreen; partridge berry; checkerberry) – an interesting creeping shrub with aromatic, medicinal, evergreen leaves, blush-white urn-shaped flowers in midsummer, followed by large red berries; good in containers. Moist acid soil in partial shade; 3–6 in (7.5–15 cm) high and 3 ft (90 cm) or so across.

x *Halimiocistus sahucii* – a spreading evergreen shrub with dark green leaves and numerous white flowers in summer. Well-drained soil in sunny sheltered position; 2½ ft (75 cm).

Hebe – useful evergreen shrubs for the small garden though many are not reliably hardy. Hebes need well-drained neutral to alkaline soil in full sun. In coastal regions they will stand exposed positions but appreciate shelter inland.

Hebe albicans – a hardy shrub with a dense rounded habit, glaucous lanceolate leaves and 2 in (5 cm) heads of white flowers in summer; 'Pewter Dome' – compact habit. 2 ft (60 cm).

Hebe armstrongii – a greenish-gold 'whipcord' hebe, tipped with small white flowers in summer; reasonably hardy; 3 ft (90 cm).

*Hebe brachysiphon** – a dense hardy shrub with dark green ovate leaves and numerous spikes of white flowers in summer; 'White Gem' – compact habit, ideal for hedging. 6 ft (1.8 m).

Hebe macrantha – a compact evergreen shrub with toothed leathery leaves and large 1 in (2.5 cm) four-petalled white flowers in summer; not reliably hardy; 2 ft (60 cm).

Hebe ochracea 'James Stirling' – a delightful miniature with gold whipcord foliage and small white flowers; excellent for containers. 9 in (23 cm).

Hebe pimeleoides 'Quicksilver' – a small semi-prostrate shrub with black twigs, silvery blue

Hebe macrantha

foliage, and mauve-tinged white flowers in summer; 12–15 in (30–38 cm) high and about 20 in (50 cm) across.

Hebe pinguifolia 'Pagei' (syn. *H. pageana*) – a hardy evergreen ground cover shrub with blue-green foliage and spikes of white flowers in early summer; 6–9 in (15–23 cm) high and about 3 ft (90 cm) across.

Hebe rakaiensis – a densely compact, rounded shrub with pale green leaves and white flowers in summer; 12–15 in (30–38 cm).

Hebe vernicosa – a compact shrub with bright green foliage and numerous white flowers in summer; 12 in (30 cm).

Helianthemum nummularium (rock rose; sun rose) – a wiry shrub with small elliptic leaves and numerous short-lived flowers in summer; 'Wisley White' – grey leaves and single white flowers. Ordinary well-drained soil in full sun; 6 in (15 cm) high and about 2 ft (60 cm) across.

Helichrysum bellidioides – a prostrate New Zealand shrublet with ovate dark green leaves, woolly beneath, and small white everlasting daisies in summer; not reliably hardy. Light, well-drained soil in a sunny sheltered position; 3 in (7.5 cm) high and about 12 in (30 cm) across.

Hibiscus syriacus – a hardy deciduous shrub with lobed leaves and large mallow flowers in summer; 'Diana' – white; 'Elegantissimus' – double white with a maroon centre; 'Wm. R. Smith' – a large crinkled white; needs a sheltered 'sun trap' in cold areas or flowering will be poor. Ordinary well-drained soil; up to 10 ft (3 m).

Hydrangea arborescens – a hardy species with flat corymbs of off-white flowers in summer; 'Annabelle' – huge pure white heads, 6–10 in (15–25 cm) across, which fade to cream; 'Grandiflora' – heavy globular pure white clusters. Rich moisture-retentive soil in partial shade, avoiding early morning sun which can damage new growths after frost; 4–6 ft (1.2–1.8 m).

Hydrangea involucrata 'Hortensis' – slightly corrugated foliage and 3–5 in (7.5–12.5 cm) heads of unusual double white florets which blush pink in sunny situations; cultivation and height as above.

Hydrangea macrophylla (common hydrangea) – 'Lanarth White' – a compact lacecap type with white florets around pink or blue centres; 'Madame Emile Mouillière' – a rather tender mop-head variety with large fringed white florets which age pink; 'White Wave' – pearly white florets, again with pink or blue centres; cultivation and height as above.

Hydrangea paniculata – a large hardy species with white flowers (ageing pink) in conical clusters, 6–8 in (15–20 cm) long, during late summer; 'Grandiflora' – panicles up to 18 in (45 cm) long; 'Kyushu'* – a tolerant variety with luxuriant bright green foliage and white panicles from midsummer to autumn; 'Praecox' – a smaller, earlier flowering shrub with 10 in (25 cm) panicles; 'Tardiva' – an erect shrub with attractively sparse panicles of cream fertile flowers interspersed with white sterile florets which are borne in late summer. Cultivation as above, though more tolerant of sun; 12–15 ft (3.5–4.5 m)

*Hydrangea quercifolia** – an unusual white-flowered hydrangea with attractive oak-leaf foliage which turns spectacular shades in autumn; 'Sterile' – large clusters of sterile white florets in summer. Cultivation in partial shade as above; 6 ft (1.8 m).

Hyssopus officinalis (hyssop) 'Albus' – narrow semi-evergreen aromatic leaves and spikes of white flowers in summer; suitable for dwarf hedging, alone or with the blue and pink forms. Well-drained soil in sun; 18 in (45 cm).

Lavandula angustifolia 'Alba'* – narrow grey-green evergreen foliage and spikes of white flowers in summer; 'Nana Alba' – broader grey-green leaves, white flower spikes and a smaller habit, reaching only 12 in (30 cm); good for dwarf hedges, in knot gardens for example. Well-drained soil in sun; 15–24 in (38–60 cm). ✕

Lavandula stoechas (French lavender) 'Alba' – aromatic grey foliage and flower spikes which are topped by showy white bracts; not reliably hardy. Cultivation as above but in a dry sheltered position; up to 2 ft (60 cm). ✕

Leptospermum humifusum – an evergreen Tasmanian species, hardier than most, which has a low, ground-covering habit, small dark green leaves, and masses of five-petalled white flowers in early summer. Well-drained soil in sun; 6 in (15 cm) high and about 3 ft (90 cm) across.

Ligustrum lucidum – a small tree with glossy dark evergreen camellia-like leaves and 6–8 in (15–20 cm) open panicles of cream flowers in late summer and autumn; needs shelter from cold winds; 'Excelsum Superbum' – cream and gold variegated foliage. Ordinary soil in sun or semi-shade; 10 ft (3 m) or more.

Ligustrum ovalifolium (privet) – a hardy evergreen Japanese shrub, commonly grown as hedging; ovate leaves and panicles of tiny, rather unpleasant smelling, tubular cream flowers (produced on untrimmed bushes), followed by black berries; 'Argenteum' (syn. 'Variegatum') cream-margined foliage – 'Aureum' (syn. 'Aureo-marginatum') – golden foliage. Ordinary soil in sun or shade; 12 ft (3.5 m) or more.

Ligustrum quihoui – an elegant shrub with purplish pubescent wiry branches and large loose panicles of white flowers in late summer. Cultivation as above; 6 ft (1.8 m).

Lonicera fragrantissima – a semi-evergreen Chinese species with ovate leaves and small, highly scented creamy-white flowers in winter. Well-drained soil in sun or partial shade; 6 ft (1.8 m).

Lonicera x *purpusii* – a deciduous shrub with ovate leaves and scented creamy-white flowers in winter; 'Winter Beauty'* – more flowers over a longer period. Cultivation and height as above.

Lonicera standishii – another winter-flowering Chinese honeysuckle, very similar in appearance to both the above; semi-evergreen, compact and free-flowering from late autumn to early spring. Cultivation as above; 4–5 ft (1.2–1.5 m).

Luma apiculata (syn. *Myrtus luma*) – a slightly tender tree with small dark evergreen leaves, piebald bark and myrtle-line flowers in late summer; 'Glanleam Gold' – cream-variegated foliage. Neutral to acid soil and a warm sheltered spot; 30 ft (9 m).

Lupinus arboreus (tree lupin) – a Californian shrub, often short-lived, with light green digitate leaves and spikes of fragrant pea flowers in summer; 'Snow Queen' – white, does well in coastal areas. Light well-drained soil in sun or partial shade; 4 ft (1.2 m).

Magnolia – the majority of magnolias have either white flowers or white-flowered forms and the following list is by no means exhaustive. Though mostly hardy in themselves, buds and blooms may sustain frost damage and therefore early-flowering magnolias should be planted in situations sheltered from frosts and cold winds. All magnolias do best in well-drained humus-rich soil, in sun or partial shade. Most do better in neutral or acid soils.

Magnolia cylindrica – a spreading shrubby tree with erect cylindrical white flowers on the bare branches in spring; 20 ft (6 m) or more.

Magnolia denudata see *M. heptapeta*.

*Magnolia grandiflora** – an American species with glossy ovate leathery, evergreen leaves, downy and rust-coloured beneath, and ivory scented cup-shaped flowers, up to 8 in (20 cm) across, in late summer; 'Exmouth' – begins to flower at an early age; 'Goliath' – shorter, broader leaves and very large globular flowers at an early age; lime-tolerant, needs the shelter of a wall in most northern temperate regions. 15 ft (4.5 m).

Magnolia heptapeta (syn. *M. denudata*) (yulan) – a slow-growing, deciduous tree with ovate leaves, downy beneath, and fragrant goblet-shaped white flowers, freely produced and 5–6 in (12.5–15 cm) across, in spring; up to 15 ft (4.5 m).

Magnolia salicifolia – an erect fast-growing deciduous Japanese species with lanceolate leaves and wide-open narrow-petalled white flowers in spring, before the new foliage; up to 20 ft (6 m).

Magnolia sieboldii – a very beautiful summer-flowering deciduous species with a spreading habit, dark green lanceolate leaves, 3 in (7.5 cm) white flowers which have a boss of crimson stamens, followed by pink seed capsules that open in early autumn to reveal orange seeds; up to 15 ft (4.5 m) high and 18 ft (5.5 m) across.

Magnolia sieboldii

Magnolia x *soulangeana* – a widely grown hybrid with deciduous foliage and goblet-shaped flowers in spring before the new leaves expand; 'Alba Superba' – fragrant pure white flowers; 'Lennei Alba' – very large, more rounded white flowers; up to 15 ft (4.5 m).

Magnolia stellata – a compact slow-growing deciduous species with starry white flowers, 3–4 in (7.5–10 cm) across, on the bare branches in spring; 'Royal Star'* – a very compact variety with numerous large flowers; 'Water Lily' – larger flowers with broader petals; 8–10 ft (2.5–3 m).

Magnolia 'Wada's Memory'* – an erect tree which is completely covered in fragrant narrow-petalled flowers which are similar to those of *M. stellata* but much larger; up to 30 ft (9 m).

Malus (crab apple) 'Golden Hornet' – a vigorous erect deciduous tree with light green ovate leaves, white flowers in spring and abundant yellow fruits which last well into winter. Well-drained soil in sun or partial shade; up to 18 ft (5.5 m). ✕

Malus hupehensis – an oriental species with ascending branches, large fragrant white flowers (pink in bud), followed by red-flushed yellow fruits. Cultivation as above; 25 ft (7.5 m) or more. ✕

Malus 'John Downe' – an erect cultivar with large white flowers in May, followed by orangey-red fruits which are especially good for preserves. Cultivation as above; up to 30 ft (9 m). ✕

Malus 'Red Sentinel' – a white-flowered, rather graceful cultivar with long-lasting dark red fruit. Cultivation as above; 10–15 ft (3–4.5 m). ✕

Malus sargentii – a small shrubby Japanese species with ovate to trilobed leaves, yellow-stamened white flowers and small bright red fruits. Cultivation as above; 6 ft (1.8 m) or more. ✕

Malus 'Yellow Siberian' – large white flowers in spring and bright yellow, persistent fruits; up to 15 ft (4.5 m). ✕

Myrtus communis (myrtle)* – aromatic, neat, evergreen ovate leaves which are offset in summer by stamen-filled white flowers and large black berries; 'Flore Pleno' – double flowers; 'Tarentina' (syn. 'Jenny Reitenbach'; 'Microphylla', 'Nana') – a compact dwarf narrow-leaved cultivar with white berries; 'Tarentina Variegata' – the same as the preceding cultivar but with ivory variegation; 'Variegata'* – a free-flowering cultivar with cream-variegated leaves which makes an excellent pot plant; not reliably hardy. Well-drained soil in a sheltered sunny position, preferably against a wall in areas subject to hard frosts; up to 10 ft (3 m) in favourable conditions outdoors, or about 3 ft (90 cm) in a pot. ✕

Myrtus luma see *Luma apiculata*

Nandina domestica (Chinese sacred bamboo) – not a true bamboo, but a relative of *Berberis*; pinnately compound evergreen leaves, reddish when new and purplish in autumn, and 6–15 in (15–38 cm) panicles of white flowers in summer, followed by long-lasting white to red fruits; 'Firepower' – a dwarf cultivar with colourful red, orange and cream foliage; 'Nana Purpurea' – compact and reddish all year. Moist well-drained humus-rich soil in sun; needs shelter from hard frosts, though damaged plants usually survive; up to 6 ft (1.8 m).

Olearia gunniana (daisy bush) – a reasonably hardy Tasmanian species with greyish leaves and small daisy flowers in 12 in (30 cm) panicles in summer; hardiest in coastal regions. Well-drained soil in a warm sunny position; 4–5 ft (1.2–1.5 m).

Olearia x *haastii** – a hardy species from New Zealand with broad ovate leaves, felted white undersides, and clusters of tiny daisies in summer. Cultivation as above; up to 8 ft (2.4 m).

Olearia ilicifolia (Maori holly) – another hardy New Zealander, this time with leathery holly-like leaves, felted beneath, and clusters of tiny scented daisy flowers in early summer. Cultivation as above; up to 10 ft (3 m).

Olearia macrodonta (New Zealand holly) – similar to the above but with larger flower clusters; 'Major' – even larger flowerheads. Cultivation as above; up to 12 ft (3.5 m).

Olearia x *scillionensis** – rather like *O. gunniana* but with more richly coloured foliage and more numerous flowers. Cultivation as above; 4–5 ft (1.2–1.5 m).

Osmanthus delavayi* – a slow-growing evergreen Chinese species with glossy ovate leaves and clusters of small scented jasmine-like flowers in spring. Well-drained soil in sun or partial shade, with shelter from cold winds; up to 8 ft (2.4 m).

Osmanthus heterophyllus – a slow-growing evergreen Japanese species with prickly leaves, variable in shape, and clusters of tiny scented tubular flowers in autumn; 'Aureomarginatus' – yellow-edged leaves; 'Gulftide' – dense, extra-spiny foliage; 'Latifolius Variegatus' – broad silver-variegated leaves; 'Pur-

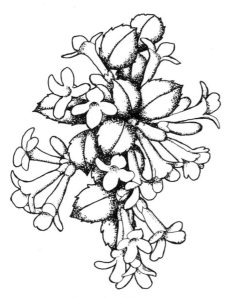

Osmanthus delavayi

pureus' – purple-flushed foliage, especially when young; 'Variegatus' – leaves with ivory margins; suitable for hedging, but cutting restricts flowering. Cultivation as above; 10 ft (3 m).

x *Osmarea burkwoodii** – a hybrid between *Osmanthus delavayi* and *Pyllyrea decora* which has shiny ovate evergreen leaves and clusters of fragrant tubular white flowers in spring; suitable for hedging. Ordinary soil in sun or partial shade; 10 ft (3 m).

Pachysandra terminalis – a creeping Japanese sub-shrub with angular evergreen leaves and spikes of white-stamened flowers in spring; 'Variegata' – a less vigorous form with white-variegated foliage; excellent ground-cover in light or heavy shade; 12 in (30 cm) high and 18 in (45 cm) across.

Paeonia suffruticosa (tree peony; moutan) – a deciduous shrub with light green divided leaves and sumptuous silky white flowers in late spring which are 6 in (15 cm) or more across and blotched purple to pink around the golden stamens; 'Flight of Cranes' (syn. 'Ren-kaku') – huge pure white semi-double flowers on a large shrub; 'Jewelled Screen' (syn. 'Tama Sudare')* – a floriferous variety with large fragrant semi-double flowers; 'Kingdom of the Moon' (syn. 'Gessekae') – a very large white with frilled petals; 'King of White Lion' (syn.

'Hakuo Jisi') – an easy free-flowering variety with purple-centred white flowers; 'Large Globe' (syn. 'Godaishu') – giant rounded flowers; 'Mrs William Kelway' – a gold-centred semi-double; 'Rock's Variety'* – single or semi-double white flowers with a maroon centre. Rich well-drained soil in sun or partial shade, with protection from early morning sun which will damage buds and flowers after frost; 3–6 ft (90 cm–1.8 m).

Parahebe catarractae 'Alba' – a miniature evergreen bush which forms spreading mounds and is covered in tiny white flowers during early summer. Well-drained soil in sun; 6 in (15 cm).

Penstemon scouleri 'Albus' – a shrubby species native to Washington State, with lanceolate leaves and tubular flowers in summer. Well-drained soil in sun; 9–12 in (23–30 cm) high and about 18 in (45 cm) across.

Philadelphus (mock orange) – easy, tolerant, deciduous shrubs for ordinary well-drained soil in sunny or semi-shaded sites.

Philadelphus 'Beauclerk'* – a hybrid mock orange with large pink-centred flowers in summer; up to 8 ft (2.4 m).

Philadelphus 'Belle Etoile'* – very fragrant maroon-centred flowers; up to 10 ft (3 m).

Philadelphus 'Burfordiensis' – single flowers with prominent yellow stamens; up to 10 ft (3 m).

Philadelphus coronarius (mock orange) – an easy, hardy floriferous deciduous shrub with ovate mid-green leaves and single white flowers which smell like orange blossom; 'Aureus' – yellow new leaves which age to lime green. The golden form scorches in full sun and retains colour best in shade with regular pruning of old wood (which should be done after flowering); 'Variegatus' – white-edged leaves; 5–9 ft (1.5–2.7 m).

Philadelphus 'Innocence' – creamy-yellow variegated leaves and abundant single scented white flowers. Height as above.

Philadelphus 'Manteau d'Hermine'* – a dwarf compact cultivar with fragrant double flowers; 3 ft (90 cm).

Philadelphus microphyllus – an American species, similar to *P. coronarius* but smaller in all its parts; 2–3 ft (60–90 cm).

Philadelphus 'Virginal' – fragrant fully double flowers; 8–9 ft (2.4–2.7 m).

Photinia x *fraseri* (Chinese hawthorn) 'Red Robin' – an evergreen shrub with bright red new leaves and white flowers in early summer; similar to *Pieris* but lime-tolerant; suitable for hedging. Further shows of new red leaves can be produced by careful pruning throughout the growing season. Well-drained soil in sun or partial shade; up to 8 ft (2.4 m).

Photinia glabra 'Rosea Marginata' – a slow-growing evergreen species with white flowers and foliage variegated pink, green, grey and white (bright pink when young); needs winter protection in cold areas when young. Cultivation as above; 3½–5 ft (1.05–1.5 m).

Physocarpus opulifolius 'Dart's Gold' – a splendid, small, hardy golden shrub, reaching barely 3 ft (90 cm); three- or five-lobed leaves and clusters of small white flowers in early summer; 'Luteus' – bright yellow foliage and larger, reaching 6 ft (1.8 m). Well-drained moisture-retentive soil in full sun.

Pieris 'Firecrest' – an evergreen hybrid, hardier than most, with leathery, dark green elliptic foliage, upright clusters of red new leaves and panicles of lovely bell-shaped white flowers in late spring. Moist lime-free soil in partial shade; about 6 ft (1.8 m).

Pieris 'Forest Flame' – an outstanding cultivar whose new red leaves turn pink and then cream before becoming dark green; flowering, cultivation and height as above, with the addition of shelter from hard or late spring frosts.

Pieris formosa var. *forrestii* – bright red new leaves and 3–5 in (7.5–12.5 cm) panicles of white bells in spring; 'Charles Michael' – large flowers; 'Wakehurst' – brilliant red new growth and glistening white flowers; may be cut back by severe frosts but usually survives. Cultivation as above; 12 ft (3.5 m).

Pieris japonica – shiny coppery-red new leaves and 3–4 in (7.5–10 cm) wide panicles of white bells in spring (earlier than *P. formosa*); 'Debutante' – a small, compact, very hardy cultivar, about 30 in (75 cm) tall and wide, with dark foliage and graceful trusses of white bells; 'Mountain Fire' – a small growing cultivar, not much more than 3 ft (90 cm) tall, with vivid new red leaves and large clusters of pure white bells; 'Purity'* – probably the best cultivar for flowers, with abundant large flower clusters and contrasting dark green leaves; 'Variegata' – a slow-growing compact cultivar, reaching only 30 in (75 cm), with narrow, cream-edged leaves which are pink-flushed when young. Cultivation as above; up to 10 ft (3 m) unless stated otherwise.

Pittosporum dallii – a rather tender evergreen New Zealand shrub with large leathery ovate leaves, red when young, and terminal clusters of scented white flowers in summer. Needs a very sheltered sunny position with well-drained sandy soil; hardier in coastal areas; may also be grown as a conservatory plant; up to 10 ft (3 m) but pot plants may be pruned after flowering to maintain a compact shape.

Poncirus trifoliata (Japanese bitter orange; hardy orange) – an oriental shrub with fearsome spines, leathery dark green trifoliate leaves, 2 in (5 cm) scented white flowers in spring, followed by small yellow bitter-tasting orange-like fruits; used for hedging in the southern United States; not reliably hardy in cold areas. Best in slightly acid soil in full sun; 10 ft (3 m) or more.

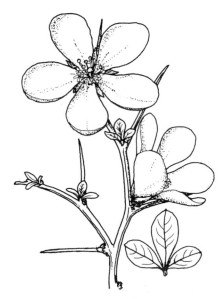

Poncirus trifoliata

Potentilla dahurica (syn. *P. fruticosa*) 'Abbotswood'* – an easy, reliable, free-flowering shrub with greyish leaves and white rock rose flowers throughout the summer; 'Farrer's White' – a white-flowered cultivar with an erect habit; 'Mandschurica' – a mat-forming cultivar, about 12 in (30 cm) high and up to 3 ft (90 cm) across, which has purple twigs, silvery foliage and white flowers; 'Tilford Cream' – a dwarf compact cultivar, reaching only 18 in (45 cm), with grey-green leaves and creamy-white flowers; 'Vilmoriniana' – an upright shrub with silver-grey leaves and ivory flowers. Well-drained soil in sun; 4–5 ft (1.2–1.5 m).

Potentilla fruticosa see *P. dahurica*.

Prunus – a large and varied group of trees and shrubs which on the whole are shallow-rooted and dislike deep planting or deep cultivation around the base. Most prefer neutral to slightly alkaline soil in a sunny position but it is advisable to check individual requirements on purchase.

Prunus amygdalus see *Prunus dulcis*.

Prunus avium (wild cherry; gean) – a vigorous deciduous tree with pendant clusters of white flowers in spring; 'Plena' – longer lasting double flowers, freely produced; up to 40 ft (12 m).

Prunus cerasifera (cherry plum) – numerous small white blossoms in late winter and early spring, followed by ovate mid-green leaves and edible fruits; 'Atropurpurea' (syn. 'Pissardii') – purple leaves and pink-tinged flower buds, opening to blush white; suitable for hedging; up to 25 ft (7.5 m). ✕

Prunus cerasus (sour cherry) 'Rhexii' – double white flowers in spring; up to 20 ft (6 m). ✕

Prunus x *cistena* (purple-leaved sand cherry) – small crimson leaves and white flowers in spring; suitable for hedging; 4 ft (1.2 m). ✕

Prunus communis see *Prunus dulcis*.

Prunus davidiana 'Alba' – an erect tree, closely related to the peach (*P. persica*) with white blossoms on the bare branches in winter and early spring, followed by pointed lanceolate leaves; needs a sheltered position; up to 30 ft (9 m). ✕

Prunus dulcis (syn. *P. amygdalus*; *P. communis*) (almond) 'Alba' – single white blooms on bare branches in spring, followed by lanceolate mid-green leaves and, in warm regions, edible almonds; up to 25 ft (7.5 m). ✕

Prunus glandulosa (Chinese bush cherry) 'Albiplena'* – in fact, a Chinese almond with a compact shrubby habit and double white flowers in spring; ideal for small gardens where a tree is impractical. Needs a sheltered site; 4 ft (1.2 m). ✕

Prunus 'Hally Jolivette'* – a graceful willowy tree with a long flowering period; abundant semi-double blush white blossoms in spring; ideal for small gardens; up to 20 ft (6 m). ✕

Prunus 'Jo-nioi'* – a Japanese hybrid cherry with a spreading habit, pale brown young leaves and fragrant white flowers in late spring; up to 25 ft (7.5 m) high and 30 ft (9 m) across. ✕

Prunus laurocerasus (cherry laurel; common laurel) – glossy evergreen leaves and 3–5 in (7.5–12.5 cm) spikes of small creamy white flowers in spring, followed by black fruits; 'Otto Luyken'* – an attractive dwarf laurel with neat narrow leaves and numerous erect flower spikes. Tolerates a wide range of conditions, including heavy shade; 3–4 ft (90 cm–1.2 m).

Prunus lusitanica (Portuguese laurel) – a hardy tolerant shrub with glossy dark green, red-stalked ovate leaves and 6–8 in (15–20 cm) racemes of scented cream flowers in early summer, followed by black fruits; up to 20 ft (6 m); 'Variegata' – a slow-growing cultivar with white-variegated leaves, sometimes flushed pink in winter, which reaches only 7 ft (2.1 m).

Prunus mume (Japanese apricot) 'Alboplena' – a small early flowering tree with dense clusters of semi-double white blossoms in winter and early spring. Needs a sheltered position; up to 25 ft (7.5 m). ✕

Prunus padus (bird cherry) – a distinctive tree with dense spikes, 3–5 in (7.5–12.5 cm) long, of almond-scented white flowers in late spring; 'Watereri' – abundant spikes up to 8 in (20 cm) long; up to 30 ft (9 m).

Prunus serrula – a Chinese species with gleaming reddish-brown bark and tiny white blossoms in spring which tend to be hidden among the new leaves; up to 25 ft (7.5 m).

Prunus 'Shirofugen' – a vigorous Japanese hybrid cherry with copper new leaves and double white blossoms in late spring which are pink in bud and fade to purplish-pink; up to 30 ft (9 m). ✂

Prunus 'Shirotae' (syn. 'Kojimae')* – a vigorous spreading Japanese cherry with horizontal to drooping branches, bronze new leaves, and single or semi-double scented white blossoms in spring; up to 25 ft (7.5 m). ✂

Prunus spinosa (blackthorn; sloe) – a dense shrub with blackish spiny branches which are covered in small white blossoms in early spring, followed by blue-black fruits; 'Plena' – longer lasting double flowers; 'Purpurea' – purple foliage; suitable for hedging; 10–15 ft (3–4.5 m).

Prunus subhirtella 'Autumnalis' (autumn cherry)* – a valuable tree which flowers when little else does, with fragile semi-double white flowers from late autumn to the end of winter; 20 ft (6 m) or more. ✂

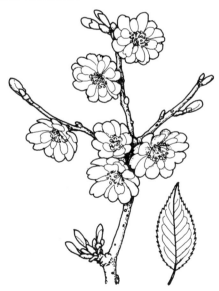

Prunus subhirtella *'Autumnalis'*

Prunus 'Tai-haku' (great white cherry) – a vigorous Japanese hybrid with coppery-red new leaves and superb clusters of large wide-open single white blossoms in spring; up to 30 ft (9 m). ✂

Prunus 'Umeniko'* – a narrow upright Japanese hybrid with freely produced white flowers in spring as the new leaves unfurl; good autumn colour; excellent for restricted spaces; up to 20 ft (6 m) high and 5–6 ft (1.5–1.8 m) wide. ✂

Prunus x *yedoensis* (Yoshino cherry) – a Japanese tree with graceful arching branches and profuse almond-scented blush white blossoms in spring; up to 25 ft (7.5 m); 'Ivensii' – a weeping variety, about 10–15 ft (3–4.5 m) tall and as much, or slightly more across, which bears scented pure white flowers. ✂

Pyracantha (firethorn) – there are over thirty different pyracanthas in cultivation and virtually all have cream or white hawthorn-like flowers but they are mainly grown for their abundant bright red, orange or yellow berries. They are lime-tolerant and easily grown in ordinary well-drained soil in sun or partial shade.

Pyracantha atalantioides – a large upright quick-growing shrub with flat clusters of white flowers in early summer, followed by dark red berries; 'Aurea' – yellow berries; up to 15 ft (4.5 m).

Pyracantha coccinea – narrow pointed leaves and clusters of white flowers, followed by red berries; 'Lalandei Monrovia' – broader leaves and profuse orange fruits; up to 15 ft (4.5 m); 'Sparkler' – a smaller spreading shrub with variegated foliage, flushed pink in winter, and red fruits.

Pyracantha crenulata 'Rodgersiana' – similar in appearance to the above, with orange-red berries; 'Flava' – yellow-berries; 8–10 ft (2.4–3 m).

Rhaphiolepis umbellata (Yeddo hawthorn; Indian hawthorn) – a dense evergreen shrub with leathery rounded dark green leaves and clusters of fragrant white flowers in late spring, followed by blue-black berries; reasonably hardy in mild areas, otherwise an excellent subject for pots in a cool conservatory. Well-drained soil with additional peat and sand in a sunny sheltered position; 4 ft (1.2 m).

Rhododendron – the majority of rhododendrons need moisture-retentive well-drained lime-free soil which has a high humus content. They prefer partial shade in positions sheltered from strong winds. The hardiness of

Rhaphiolepis umbellata

individual species and varieties should be determined on purchase. The term 'azalea' refers to various (mostly deciduous) species and their hybrids.

Rhododendron arborescens – a hardy deciduous North American azalea with fragrant white flowers in summer; 15 ft (4.5 m).

Rhododendron arboreum – an evergreen tree with large long leaves, creased on the upper surface and felted brown underneath, whose white-flowered forms are reasonably hardy (unlike the red, which are rather tender); 'Cinnamomeum' – gingery-brown leaf undersides and white flowers in spring; 'Sir Charles Lemon' – rust-coloured undersides and white flowers in late winter and early spring; 20–40 ft (6–12 m).

Rhododendron 'Arctic Tern'* – a hardy dwarf shrub with pure white flowers in early summer; 2 ft (60 cm).

Rhododendron atlanticum – a North American deciduous suckering species with grey foliage and narrow, intensely fragrant white flowers which are flushed pink and sticky on the outside; 3 ft (90 cm).

Rhododendron auriculatum – an evergreen species for patient gardeners, with leaves up to 12 in (30 cm) long and large trusses of fragrant white flowers; midsummer flowering, and only on mature plants. Needs shelter from strong winds which tear the large leaves; 10–15 ft (3–4.5 m).

Rhododendron calophytum – another long-leaved species, with drooping narrow foliage like a skirt beneath the huge clusters of white bell-shaped flowers which have reddish stalks and basal blotches; best chosen when flowering as some plants bear pink flowers; 10–15 ft (3–4.5 m).

Rhododendron crassum – an evergreen tree or shrub with pointed leaves, creased on the upper surface and rust-coloured underneath, and small clusters of huge fragrant white flowers in summer. Not hardy in cold areas; up to 15 ft (4.5 m).

Rhododendron decorum – an evergreen species with grey-green leaves which have bluish undersurfaces, and large fragrant green-centred white flowers in spring (though some plants have pale pink flowers); a reasonably hardy and fairly small species for the gardener who values fragrance highly but has limited space. Avoid sites which get early morning sun; 6–10 ft (1.8–3 m).

Rhododendron 'Dora Amateis'* – a small compact evergreen shrub with profuse clusters of pure white flowers in late spring; 2 ft (60 cm).

Rhododendron fictolacteum – an evergreen tree with long leathery leaves, felted pale brown on new growths and undersurfaces, and large clusters of bell-shaped white flowers which are blotched crimson inside. Needs shelter from strong wind; up to 20 ft (6 m).

Rhododendron flavidum 'Album' – a small evergreen shrub with neat shiny leaves and white flowers in early spring; up to 5 ft (1.5 m).

*Rhododendron griffithianum** – the grand master of the fragrant whites; a magnificent evergreen species with 12 in (30 cm) leaves, blue-grey underneath, and trusses of four to six massive scented white bells, some 5–6 in (12.5–15 cm) across at the mouth and 3 in (7.5 cm) deep; late spring flowering, and only on mature plants. Needs a very genial climate to succeed; up to 20 ft (6 m).

Rhododendron johnstoneaum – an evergreen species with creamy-white scented flowers mid-season; 'Flora Pleno' – the double-flowered form; 15 ft (4.5 m).

Rhododendron 'Kure-no-yuki' – an unusual evergreen Japanese azalea hybrid with white semi-double hose-in-hose flowers; 3–4 ft (90 cm–1.2 m).

Rhododendron leucaspis – a dwarf evergreen species with ovate leathery greyish leaves and 2 in (5 cm) brown-stamened white flowers with wavy petals which are borne in ones and twos from late winter to spring. Needs protection from frost if the flowers are to remain undamaged; may be grown as a pot plant in a frost-free greenhouse; 12–24 in (30–60 cm).

Rhododendron mucronatum – a spreading azalea with semi-evergreen, slightly hairy, bright green leaves and pure white flowers in late spring; about 4 ft (1.2 m).

Rhododendron 'Palestrina'* – a hardy evergreen hybrid azalea with pure white flowers, delicately striped green; 3–4 ft (90 cm–1.2 m).

Rhododendron quinquefolium – a deciduous azalea with rounded purple-flushed leaves and white green-throated flowers in spring; good autumn colour; up to 5 ft (1.5 m).

Rhododendron 'Snow Queen' – an evergreen hybrid with pink-flushed buds opening to large trumpet-shaped flowers with red basal blotches in late spring; 15–20 ft (4.5–6 m).

Rhododendron viscosum (swamp honeysuckle) – a completely hardy deciduous azalea with light green obovate leaves, furry underneath, and clusters of small fragrant white flowers which in some plants are pink-flushed; 6–8 ft (1.8–2.4 m).

Rhododendron 'White Glory' – a large-flowered evergreen hybrid with cream buds opening to white in early spring; about 10 ft (3 m).

Rhododendron 'Whitethroat' – a deciduous azalea hybrid with pure white frilly double flowers in late spring and good autumn colour; 6 ft (1.8 m).

*Rhododendron yakushimanum** – a lovely dwarf evergreen shrub, ideal for containers; silvery felted new growths, leathery leaves with curled-under margins and woolly beige undersides, and pink buds which open to white bell-shaped flowers in late spring; 2 ft (60 cm) high and up to 3 ft (90 cm) across.

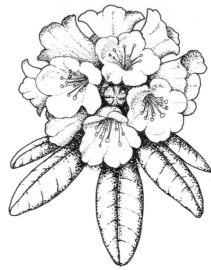

Rhododendron yakushimanum

Ribes sanguineum (flowering currant) 'Tydeman's White'* – a hardy deciduous shrub with drooping grape-like clusters of ivory flowers in spring and soft trilobed leaves. Well-drained soil in sun or partial shade; 6 ft (1.8 m). ✂

Romneya coulteri (tree poppy)* – a slightly tender summer-flowering sub-shrub from California, which spreads by underground runners and bears blue-green, deeply cut leaves and 4–5 in (10–12.5 cm) flat white crepe poppies with a tight boss of golden stamens; resents disturbance. Light well-drained soil in a sunny sheltered position; needs winter protection in cold areas; up to 8 ft (2.4 m).

*Romneya trichocalyx** – very similar to the above, but smaller, more erect, and slower growing, though spreading further eventually. Cultivation as above; 3 ft (90 cm).

Rosa (rose) – roses do best in well-drained moisture-retentive soil in a light, open position. They prefer deep, slightly acid soil but are lime-tolerant. Even unpromising conditions, such as shallow soil over chalk or dry sandy soil, may be improved sufficiently for roses to succeed by the addition of peat and well-rotted organic matter. Albas, gallicas and damasks cope best with poor soil.

Rosa x *alba* (white rose of York) – a very old hybrid and a most beautiful shrub, with grey-green leaves and small clusters of very fragrant semi-double flowers in early summer, followed by elongated red hips; 'Maxima' (Jacobite rose)* – rather untidy double white blooms, blush when first open; 'Semi-Plena' – heavily scented, near-single white flowers; 6ft (1.8m).

Rosa 'Alfred Carrière'* – a vigorous reliable old climber with very scented noisette (cupped) blooms over a long period; 20ft (6m). ✂

Rosa banksiae (Banksian rose) – a vigorous, semi-evergreen, rather tender climber with clusters of small violet-scented double white flowers in early summer; needs a sunny sheltered wall in cold areas; 20ft (6m).

Rosa 'Blanche Moreau' – a slender moss rose with brownish mossy glands and creamy-white fragrant double flowers in summer; 6ft (1.8m) tall and 4ft (1.2m) across. ✂

Rosa 'Bobbie James' – a wichuraiana rambler with rich foliage and massive heads of creamy-white semi-double fragrant blooms; excellent for growing into trees; 30ft (9m).

Rosa 'Boule de Neige' – a hardy vigorous Bourbon with richly fragrant double white globular flowers, crimson in bud; blooms from early summer until autumn frosts; suitable for arches and pergolas, 6ft (1.8m). ✂

Rosa bracteata (Macartney rose) – a slightly tender, semi-evergreen climber with very thorny stems and a long succession of lemon-scented single white flowers with rich gold stamens until the first autumn frosts; needs a warm, sheltered wall and is ideal for large courtyards and patios; up to 22ft (7m).

Rosa brunonii (summer-flowering musk rose) – a rampant, rather tender climber with long greyish leaves, large clusters of small musk-scented single flowers in summer, and small oval reddish-brown hips; 'La Mortola' – the same but hardier; needs a warm sheltered wall; 30ft (9m) or more.

Rosa centifolia 'Muscosa Alba' (syn. 'White Bath')* – a white-flowered sport of the moss rose; intensely fragrant 3in (7.5cm) double blooms in summer; 3–5ft (90–150cm). ✂

Rosa damascena (damask rose) – an Asian species with grey-green leaves and fragrant double flowers which are pink in bud and fade to blush-white; 6ft (1.8m). ✂

Rosa 'Elegant Pearl' – a compact 'patio floribunda' with single creamy-white flowers all summer; 2–3ft (60–90cm).

Rosa 'Félicité et Perpétué' – a vigorous hardy evergreen rambler with masses of small, lightly fragrant creamy-white flowers which are pink in bud; 20ft (6m).

Rosa filipes 'Kiftsgate'* – a rampant climber with grey-green foliage, coppery when young, and enormous panicles of small single scented white flowers in midsummer; ideal for covering extensive banks or climbing into large trees; 40ft (12m) or more.

Rosa 'Frau Karl Druschki' – an old, very elegant but scentless hybrid tea with beautifully shaped pure white buds and blooms that shade to greenish-yellow in the centre; 4ft (1.2m). ✂

Rosa gentiliana – a hardy climber with glossy foliage, large trusses of orange-scented white flowers which are palest yellow in bud, and long-lasting scarlet hips; 16ft (4.9m).

Rosa gigantea – a rather tender climber, native to Burma and southwest China, which has dense shiny dark green foliage, creamy white single flowers up to 8in (20cm) across in early summer, followed occasionally by elongated orange hips; 30ft (9m) or more.

Rosa helenae – a vigorous but rather tender scrambling climber with hooked thorns, large clusters of small, strongly scented single creamy-white flowers in midsummer, followed by profuse orange-red hips; needs a sheltered sunny position; 20ft (6m).

Rosa 'Iceberg'* – a tall vigorous floribunda with large clusters of delicately fragrant, well-shaped flowers at regular intervals until autumn; 5ft (1.5m); 'Climbing Iceberg' – the climbing version which reaches 12ft (3.5m); one of the best modern white roses, in spite of a tendency to mildew and black spot. ✂

Rosa laevigata (Cherokee rose) – a semi-evergreen Chinese species, now naturalised in the southern United States, which has handsome glossy foliage and finely scented 4in (10cm) creamy-white single flowers in early summer; needs a very sheltered site; 20ft (6m).

Rosa 'Little White Pet' – a dwarf sport of 'Félicité et Perpétué' with lightly fragrant, fully double rounded blooms; good ground-cover; 2ft (60cm) high and 2½ft (75cm) across.

Rosa longicuspis – a vigorous semi-evergreen climber with glossy foliage and large panicles of banana-scented 2in (5cm) single white flowers in early summer, followed by oval scarlet hips; up to 25ft (7.5m).

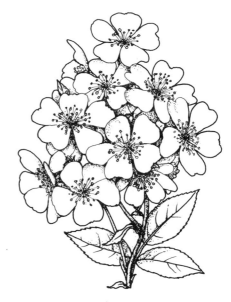

Rosa longicuspis

Rosa 'Madame Hardy'* – a perfectly shaped white damask rose, summer-flowering and fragrant, with tightly double, flat four-quartered flowers, blushed pink in bud, and broad, grey-green leaves; 6ft (1.8m). ✂

Rosa 'Moonlight' – a perpetual flowering hybrid musk with huge clusters of exotically fragrant semi-double creamy-white blooms; 6ft (1.8m). ✂

Rosa moschata (musk rose) – an autumn-flowering shrub with small clusters of 2in (5cm) single creamy-white musk-scented flowers in late summer and autumn; 12ft (3.5m). ✂

Rosa multiflora – a robust climber with large pyramidal clusters of small white flowers in summer followed by small round red hips; 10–15ft (3–7.5m).

Rosa 'Nevada'* – a recurrent-flowering shrub rose with magnificent 4 in (10 cm) semi-double creamy-white blooms; remove old wood periodically to maintain vigour; 7 ft (2.1 m).

Rosa 'Nyveldt's White' – like a white rugosa but with longer, paler leaves and more pointed flowers; 5 ft (1.5 m).

Rosa 'Pascali' – a tall upright hybrid tea with freely produced, fragrant white flowers; 4 ft (1.2 m). ✕

Rosa x *paulii* – a thicket-forming shrub with clove-scented single 2½ in (6 cm) white flowers throughout the summer; makes impenetrable ground-cover for banks and may be trained as a climber over tree stumps; 4 ft (1.2 m) high and up to 15 ft (4.5 m) across.

Rosa 'Paul's Lemon Pillar'* – a strong climber and one of the loveliest white roses in existence with few but superbly scented ivory flowers which shade to yellowish-green in the centre and have beautifully formed buds and waxy, curled-back petals; 20 ft (6 m). ✕

Rosa 'Pax' – a hybrid musk with a robust arching habit and numerous large near-single scented flowers, creamy-white shading to yellow in the centre; 6 ft (1.8 m). ✕

Rosa pimpinellifolia (syn. *R. spinosissima*) (Scotch rose; burnet rose) – a variable species with dense prickly bristles, ferny foliage and creamy-white flowers in early summer, followed by small round purplish-black hips; 'Double White' – a good double form; 'Nana' – small, almost double white flowers; up to 3 ft (90 cm) high and about 4 ft (1.2 m) across.

Rosa 'Polar Star'* – one of the best white hybrid teas; Rose of the Year 1985; 4 ft (1.2 m). ✕

Rosa rugosa 'Alba'* – a vigorous shrub with bristly stems, glossy foliage and fragrant single white flowers, 3 in (7.5 cm) across and blush white in bud, from early summer until autumn, followed by large round red hips and bright yellow autumn colour; 'Blanc Double de Coubert'* – semi-double; good for hedging; 7 ft (2.1 m) high and 4 ft (1.2 m) across.

Rosa 'Sally Holmes' – a compact shrub rose with large clusters of creamy-white blooms; 4 ft (1.2 m). ✕

Rosa 'Sander's White Rambler' – a vigorous rambler for arches and trellises with large clusters of scented pure white double flowers; 15 ft (4.5 m).

Rosa 'Schneezwerg' – a continuous-flowering hybrid rugosa with small flat semi-double pure white blooms, followed by showy orange hips; excellent for hedging; 5 ft (1.5 m).

Rosa 'Seagull' – a multiflora rambler with grey-green foliage and large trusses of small fragrant gold-stamened white flowers; good for climbing into trees; 15 ft (4.5 m).

Rosa sericea pteracantha – a distinctive shrub with large, dark red, translucent thorns, ferny foliage and little four-petalled white flowers; 8 ft (2.4 m).

Rosa 'Silver Moon' – a laevigata hybrid with luxuriant shiny foliage and large single creamy-white flowers which are apple-scented; up to 30 ft (9 m).

Rosa 'Snow Carpet' – a ground-covering shrub with miniature leaves and double white flowers; 18 in (45 cm) high and 3 ft (90 cm) across.

Rose '*Snow Carpet*'

Rosa spinosissima see *Rosa pimpinellifolia*.

Rosa 'Stanwell Perpetual' – a long-flowering bush rose with grey-green foliage and large flat semi-double blush-white blooms; ideal for hedging; 4 ft (1.2 m). ✕

Rosa 'Swan Lake' – a repeat-flowering climber with pink-centred double white hybrid tea blooms which suffer little or no petal browning in wet weather; 8 ft (2.4 m). ✕

Rosa 'The Garland' – a vigorous climber with a profusion of semi-double creamy-white, fruit-scented flowers which are buff-tinted and narrow-petalled; may be grown into a tree or as a sprawling bush; 15 ft (4.5 m).

Rosa 'Wedding Day' – a rampant climber with massive clusters of fragrant white flowers which are salmon in bud and yellowish when first open; 30 ft (9 m).

Rosa 'White Cockade' – a perpetual flowering semi-climber with glossy, disease-resistant foliage and fragrant white blooms; ideal for restricted spaces against pillars and low fences etc.; 7 ft (2.1 m). ✕

Rosa 'White Wings' – a slow-growing hybrid tea with long buds and large pure white flowers with distinctive crimson anthers; 2½ ft (75 cm). ✕

Rosa wichuraiana – a vigorous hardy semi-evergreen climber or scrambler with clusters of gold-stamened, fragrant single white flowers, up to 1½ in (4 cm) across, in late summer; suitable for ground-cover or growing into trees; 20 ft (6 m).

Rosa 'Yvonne Rabier' – a bushy floriferous polyantha with fragrant double white flowers; 4 ft (1.2 m). ✕

Rosmarinus officinalis (rosemary) 'Albus' – a white-flowered form of the aromatic herb; an attractive foliage plant with small narrow leathery dark evergreen leaves, sprinkled with white flowers in early spring; not reliably hardy in cold or very wet winters. Well-drained soil in full sun; 3 ft (90 cm) or more.

Rubus calycinoides – a prostrate creeping evergreen shrub with glossy foliage, felted white underneath, and white bramble flowers in summer. Well-drained soil in sun or partial shade; 6 in (15 cm) high, mat-forming.

Rubus deliciosus – a deciduous thornless shrub with light green, toothed palmate leaves and white dog rose flowers, up to 2 in (5 cm) across, in early summer; cultivation as above; 6–10 ft (1.8–3 m).

Rubus x *tridel* 'Benenden' – a deciduous arching shrub with trilobed leaves on bristly stems and lovely white dog rose flowers in late spring; 6–8 ft (1.8–2.4 m).

Sambucus canadensis (American elder) – a hardy deciduous shrub with pinnate leaves and flat-topped umbels of white flowers in summer, followed by purplish-black berries; 'Aurea' – golden leaves and red fruits; 'Maxima' – larger leaves and flowerheads, both reaching up to 18 in (45 cm); moisture-retentive soil in sun or shade; 10–12 ft (3–3.5 m).

Sambucus nigra (common elder; black elder) – a deciduous shrub with pinnate mid-green leaves and umbels of scented creamy-white, pollen-laden flowers in early summer, followed by round black berries; 'Aurea' – golden-leaved; 'Aureomarginata' – irregular yellow margins; – 'Laciniata' (fern-leaved elder) – deeply dissected leaves; 'Marginata' (syn. 'Albomarginata'; 'Albovariegata') – white-variegated foliage; 'Pulverulenta' – white-speckled leaves; 'Purpurea' (syn. 'Guincho Purple') – dark purplish-brown leaves and pink-stamened flowers; cultivation as above; 10–15 ft (3–4.5 m).

Sambucus racemosa (red elder) – a hardy deciduous shrub with slender pinnate leaves and yellowish-white flowers in spring, followed by red berries in summer; 'Plumosa Aurea' – the form most commonly grown, which is slow-growing and has finely cut golden foliage, bronze-flushed when new; 'Sutherland Gold' – a larger and more robust version of the former, with foliage less likely to scorch in sun. Cultivation as above; 8–10 ft (2.4–3 m).

Santolina neapolitana (cotton lavender) 'Edward Bowles' – a dwarf evergreen shrub with fine grey foliage and cream button flowers in summer; good for dwarf hedges. Well-drained soil in sun; 2 ft (60 cm).

Sarcococca confusa (Christmas box) – a small dense evergreen shrub with glossy pointed dark leaves and spidery, very fragrant ivory flowers in winter, followed by shiny black

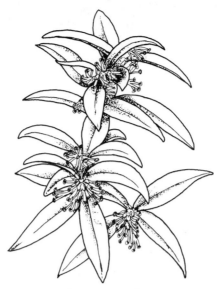

Sarcococca confusa

berries. Moisture-retentive soil in semi-shade; 2 ft (60 cm).

Sarcococca hookerana – a small dense erect evergreen shrub with narrow dark green leaves and small scented off-white flowers in winter; 'Digyna' – a more slender plant with narrower leaves; 'Purple Stem' – new growths purplish. Cultivation as above; 2 ft (60 cm).

Sarcoccoca humilis – a dwarf suckering evergreen shrub with glossy lanceolate leaves, small blush-white scented flowers in late winter, followed by black berries; good groundcover in sun or shade. Cultivation as above; 12–18 in (30–45 cm).

Sarcococca ruscifolia – a slow-growing dwarf evergreen shrub with scented white flowers in autumn, followed by dark red berries; 'Chinensis' – longer, more slender leaves and a more vigorous habit. Cultivation as above; 2 ft (60 cm).

Skimmia japonica – a hardy evergreen slow-growing shrub with dark green leathery ovate to lanceolate leaves and panicles of starry fragrant ivory flowers in late spring, followed by almost globular bright red berries; male and female plants needed for fruiting; 'Foremanii' – a female form with obovate leaves, 2 in (5 cm) panicles of fragrant cream flowers and persistent red berries; 'Fragrans' – a male form with white flowers scented like lily-of-

the-valley; 'Nymans' – a female form with a profusion of rather larger berries; 'Reevesiana' – a compact hermaphrodite which produces berries without a partner, needs lime-free soil; 'Rogersii' – a slow-growing compact female form with large fruits; 'Rogersii Nana' – a smaller-leaved, free-flowering male form; 'Rubella' – a male form with large clusters of dark red buds, opening to white flowers, in spring. Well-drained soil in sun or partial shade; tolerant of atmospheric pollution and well-suited to roadside positions; about 3 ft (90 cm), varying according to cultivar.

Sophora japonica (Japanese pagoda tree) – a hardy deciduous tree with pinnate leaves and large clusters, 6–10 in (15–25 cm) long, of creamy-white pea-like flowers in late summer or early autumn. Well-drained soil in a sheltered sunny position; 12–18 ft (3.5–5.5 m).

Sorbaria aitchinsonii – a quick-growing hardy deciduous shrub with red-stalked slender pinnate leaves and 18 in (45 cm) plumes of tiny ivory flowers in late summer. Well-drained soil in sun or light shade; 6–9 ft (1.8–2.7 m).

Sorbaria arborea – a quick-growing hardy deciduous shrub with pinnate leaves, toothed at the margin and downy beneath, and 12 in (30 cm) plumes of tiny ivory flowers in midsummer. Cultivation as above; 15 ft (4.5 m).

Sorbus (rowan; mountain ash): easy and hardy deciduous trees and shrubs, many of which are suitable for small to average-sized gardens. Most species and varieties have attractive pinnate leaves, clusters of tiny cream to white flowers, colourful (mostly red) berries and good autumn colour. They are undemanding in well-drained soil in sun or partial shade. Being grown mainly for their berries, the following list includes just a few with unusual features such as silver leaves or pale fruits.

Sorbus aria (whitebeam) – one of the few species with ovate leaves which are pearly green and downy when young, ageing to green on the upper surface and greyish underneath, with hawthorn-like panicles of cream flowers in early summer, followed by loose clusters of globular orange-red fruits and rich colour in autumn; 'Decaisneana' (syn. 'Majestica') – larger all round; 'Lutescens' – smaller and more erect, with silvery-white new leaves, turning grey-green when mature; up to 20 ft (6 m).

Sorbus forrestii* – a small tree with cream flowers followed by white berries; 15–18 ft (4.5–5.5 m).

Sorbus hupehensis – a small tree with grey-green foliage, 3 in (7.5 cm) clusters of white flowers in early summer, pink-flushed white berries and bright autumn colour; 'Obtusa' – pink fruits; up to 20 ft (6 m).

Sorbus 'Joseph Rock' – clusters of cream flowers in late spring, pale yellow to amber fruits and vivid autumn colour; up to 18 ft (5.5 m).

Sorbus koehneana – a shrub with dark green foliage, white flowers, white berries and good autumn colour; 'Harry Smith'* – smaller growing, with a heavier crop of berries; 6–8 ft (1.8–2 m).

Sorbus 'Pearly King' – white flowers and pink fruits which age to pink-flushed white; 15 ft (4.5 m).

Sorbus vilmorinii – a most elegant small tree or medium-sized shrub with immaculate fern-like foliage, 3 in (7.5 cm) clusters of white flowers in early summer, pink fruits and rich autumn colour; 7–12 ft (2.1–3.6 m).

Sorbus 'White Wax'* – a small tree with white flowers and heavy drooping clusters of white berries; up to 18 ft (5.5 m).

Spiraea x *arguta* (bridal wreath; foam of May) – a hardy deciduous shrub with slender arching branches which are covered with small white five-petalled flowers in spring, resembling the trailing fronds of a bouquet. Ordinary soil in sun; 6–8 ft (1.8–2.4 m).

Spiraea betulifolia 'Aemeliana' – a dwarf dense deciduous shrub with rounded leaves, corymbs of white flowers in spring and brilliant autumn colour. Cultivation as above; 2–2½ ft (60–75 cm).

Spiraea x *cinerea* 'Grefsheim' – narrow leaves, grey and downy when young, and small white flowers along the branches in spring. Cultivation as above; 5 ft (1.5 m) or more.

Spiraea nipponica tosaensis 'Snowmound' – a compact deciduous shrub covered in white flowers during early summer. Cultivation as above; 3 ft (90 cm).

Spiraea thunbergii – elegant white flowers along slender arching branches in spring, followed by fresh light green foliage. Cultivation as above; 5–6 ft (1.5–1.8 m).

Staphylea colchica (bladder nut) – a deciduous shrub with divided leaves and panicles of white flowers in late spring, followed by inflated fruits. Moisture-retentive soil in sun; 10 ft (3 m).

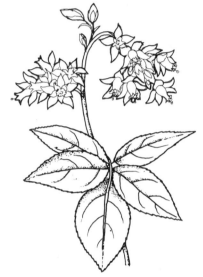

Staphylea colchica

Stewartia ovata* – a slow-growing shrub with ovate leaves, downy underneath, and 4 in (10 cm) gold-stamened ivory flowers, rather like camellias, in summer. Well-drained neutral to acid soil in semi-shade; 8–15 ft (2.4–4 m).

Stewartia pseudocamellia* – a small tree with ornamental bark, ovate leaves and 2 in (5 cm) cup-shaped white flowers in summer. Cultivation as above; 15–20 ft (4.5–6 m).

Stewartia sinensis* – a small tree with bright green ovate leaves which turn red in autumn, and scented 2 in (5 cm) white flowers in summer. Cultivation as above; 15–20 ft (4.5–6 m).

Stranvaesia davidiana – a hardy evergreen shrub or small tree with lanceolate leaves, the oldest of which turn red in autumn, and clusters of white hawthorn-like flowers in early summer, followed by globular crimson berries; 12–18 ft (3.6–5.5 m); 'Fructuluteo' –

smaller, with yellow berries, 6 ft (1.8 m); 'Pallette' – multicoloured leaves of pink, cream, orange and green; 6 ft (1.8 m). Well-drained soil in sun or partial shade.

Styrax hemsleyana (snowbell)* – a deciduous shrub or small tree with rounded leaves and racemes of white bell-shaped flowers in early summer. Well-drained moisture-retentive soil in a sheltered sunny position; 20–30 ft (6–9 m).

Styrax japonica* – a small tree with tier-shaped branches and waxy pendant white flowers in early summer; up to 25 ft (7.5 m).

Styrax obassia* – a slower growing species with rounded leaves, velvety underneath, and graceful racemes of scented white bells in summer; 20 ft (6 m) or more.

Symphoricarpos albus (snowberry) – a vigorous suckering shrub with deciduous ovate leaves and large white berries which last from autumn to the end of winter, earning it a place in the white garden despite its pink (but insignificant) flowers. Any soil in sun or shade, even under trees; 5–8 ft (1.5–2.5 m).

Symphoricarpos x *doorenbosii* 'White Hedger' – similar to the above but with smaller berries in erect clusters and a more upright, compact habit with limited suckering. Plant 2 ft (60 cm) apart for hedging, and trim two or three times each summer when established. Cultivation as above; 5–6 ft (1.5–2 m).

Symplocos paniculata – a hardy deciduous shrub with a dense twiggy habit, ovate downy leaves and terminal clusters of fragrant starry white flowers in late spring, followed by striking globular blue berries. Lime-free soil in sun; 6–8 ft (1.8–2.4 m).

Syringa x *persica* (Persian lilac) 'Alba' – a rounded deciduous shrub with lanceolate, occasionally lobed leaves, and erect pyramidal clusters of fragrant white flowers in late spring. Well-drained soil in sun; 6–8 ft (1.8–2.4 m).

Syringa vulgaris (common lilac) – a large deciduous shrub or small tree with ovate leaves and erect panicles of heavily scented flowers in late spring and early summer; 'Candeur' – cream buds and large white flowers; 'Madame Lemoine' – cream buds and double white flowers; 'Maud Notcutt'* – a vigorous white-flowered cultivar with large flowers;

'Souvenir de Alice Harding' – a late-flowering cultivar with heavily scented double ivory blooms; 'Vestale'* – a compact shrub with long loose clusters of pure white flowers. Cultivation as above; 8–12 ft (2.4–3.6 m). ✂

Viburnum – the majority of viburnums are easy in deep moisture-retentive soil in sun. Those which flower in winter and early spring should be planted in positions which avoid early morning sun after frost.

Viburnum betulifolium – a deciduous shrub with broad ovate leaves, 2–4 in (5–10 cm) corymbs of white flowers in late spring and early summer, followed by clusters of red currant-like fruits; 8–12 ft (2.4–3.6 m).

Viburnum x *bodnantense* 'Deben'* – an upright deciduous shrub with very fragrant white flowers, pink in bud, from autumn through to spring in mild areas; one of the finest, most frost-resistant winter-flowering shrubs; 9–12 ft (2.7–3.6 m).

Viburnum x *burkwoodii** – a spreading evergreen shrub with ovate leaves and clusters of deliciously scented white flowers, pink in bud, from early to late spring; 8 ft (2.4 m).

Viburnum *x* burkwoodii

Viburnum x *carlcephalum** – a deciduous shrub with broad ovate leaves and rounded clusters, 3–4 in (7.5–10 cm) across, of scented white flowers, pink in bud, during the spring; 8 ft (2.4 m).

Viburnum carlesii – a deciduous species with rather rough broadly ovate leaves and very fragrant rounded heads of white flowers, pink in bud, during the spring; 4–5 ft (1.2–1.5 m).

Viburnum farreri (syn. *V. fragrans*) – an upright deciduous shrub with bright green toothed ovate leaves and clusters of scented blush flowers, pink in bud, from autumn until spring; 'Candidissimum'* – green buds and pure white flowers; 9–12 ft (2.7–3.6 m).

Viburnum fragrans see *Viburnum farreri*.

Viburnum opulus (guelder rose) – an upright deciduous shrub with maple-like leaves, flat heads of tiny scented flowers surrounded by showy white sterile florets, followed by translucent ovoid red berries and good autumn colour; 'Aureum' – golden foliage; 'Compactum' – reaching only 6 ft (1.8 m); 'Notcutt's Variety' – larger flowerheads and fruits; 'Sterile' (snowball bush)* – ball-shaped clusters of sterile flowers (and of course no berries); 'Xanthocarpum' – yellow fruits; 15 ft (4.5 m).

Viburnum plicatum (Japanese snowball) – a deciduous shrub with white flowerheads in early summer; 'Cascade'* – layer after layer of flat white flowerheads in early summer; 'Lanarth' – more upright in habit; 'Mariesii'* – an outstanding horizontal-branched cultivar; 'Rowallane' – a dwarf upright cultivar; var. *tomentosum* – horizontally tiered branches and flattened hydrangea-like heads of ivory flowers in late spring and early summer; 'Watanabe' – a small upright compact cultivar with large flowerheads from early summer until autumn. 8–10 ft (2.4–3 m).

Viburnum rhytidophyllum – a spreading evergreen shrub with shiny corrugated lanceolate leaves, flat creamy-white flowerheads in late spring, followed by red ovoid berries which become black when ripe; fruits best when several plants are grown together; 10–15 ft (3–4.5 m).

Viburnum tinus (laurustinus) – a useful bushy evergreen species with shiny lanceolate to ovate leaves and flat heads of white flowers,

pink in bud, from autumn to spring, followed by blue to black berries; 'French White' and 'Israel'* – flowers whiter than usual; 'Lucidum' – spring-flowering, with larger leaves and flowerheads; 'Variegatum' – leaves marked with creamy-yellow; 7–10 ft (2.1–3 m).

Vitex agnus-castus (chaste tree) 'Silver Spire' – a near-hardy evergreen shrub with compound leaves and delicate spikes of tiny fragrant white flowers in early autumn. Well-drained soil in a sheltered sunny position; up to 10 ft (3 m).

Weigela 'Candida' – a compact hardy deciduous shrub with ovate pointed leaves and clusters of large white tubular flowers in early summer; 'Mont Blanc' – a hardy deciduous hybrid with very large white fragrant tubular flowers; 'Nivea' – large fragrant white flowers and leaves with felted white undersurfaces. Well-drained soil in sun or partial shade; 5–6 ft (1.5–1.8 m).

Xanthoceras sorbifolium – a hardy deciduous shrub or small tree with rowan-like leaves and 4–8 in (10–20 cm) racemes of recurved five-petalled flowers in late spring, yellow in the centre when first open, changing to purplish-pink as they age, followed by 2 in (5 cm) pendant fruits which split to reveal large brown seeds. Well-drained to dry soil in a warm sunny position; 20 ft (6 m).

Xanthoceras sorbifolium

Tender Shrubs

Abutilon x *hybridum* – a densely branched evergreen shrub with palmate leaves and pendant bell-shaped flowers; an easy subject for pots in the conservatory, best positioned where you can look up into the stamen-studded lanterns; 'Boule de Neige' – pure white flowers with orange stamens. John Innes No. 2; feed well during the growing season and keep a look out for white fly; overwinter at a minimum temperature of 50°F (10°C) and cut back in the spring; 2–4 ft (60 cm–1.2 m) as a pot plant 6 ft (1.8 m) or more in border.

Argyranthemum frutescens (syn. *Chrysanthemum frutescens*) (marguerite; Paris daisy) – an evergreen slightly tender perennial or sub-shrub from the Canary Islands which flowers for much of the year in the cool greenhouse; variable foliage from fresh to blue-green, and perfect yellow-centred white daisies. John Innes No. 2 in sun; lower growths may be removed to form a standard; up to 3 ft (90 cm). ✕

Bouvardia longiflora – a Mexican species with ovate shiny evergreen leaves and 4 in (10 cm) clusters of fragrant tubular white flowers in autumn and early winter. John Innes No. 2; ample warmth and humidity, with light shade during summer, pinching out growths several times to encourage bushiness; keep on the dry side in full sun and at a minimum of 45°F (7°C) in winter; 3 ft (90 cm).

Brunfelsia americana (lady-of-the-night) – an evergreen West Indian species with night-scented cream to white flowers, lovely for evening moments in the conservatory. Peat-based compost; warmth, ample moisture, good ventilation and light shade in summer; on the dry side at a minimum of 50°F (10°C) in winter; cut back a little in early spring and pinch out new shoots once to encourage bushiness; 4–8 ft (1.2–2.4 m).

Brunfelsia undulata – an upright evergreen shrub with lanceolate leaves and sweetly scented white to cream flowers in summer and autumn. Cultivation as above; up to 4 ft (1.2 m).

Buddleia asiatica – a compact graceful Indian species with finely serrated lanceolate leaves and honey-scented white flowers in long dense racemes during late autumn. John Innes No. 3 in full sun, kept frost-free in winter; 3–4 ft (90 cm–1.2 m).

Camellia sasanqua – a bushy species with dark evergreen elliptic leaves and generally single 1–2 in (2.5–5 cm) flowers in winter; 'Duff Alan' – single white; 'Narumi-Gata' – single scented white flowers, pink in bud. Needs full sun in winter for successful flowering and therefore best against a conservatory wall; lime-free compost; 10 ft (6 m) or more. ✕

NOTE: All white-flowered camellias are more sensitive to frost than reds and therefore are best grown under glass in areas subject to frosts and cold winter winds. The main list of white camellias is on page 133. The early-flowering *C. japonica* 'Nobilissima' and rather tender 'Mathotiana Alba' are certainly less risky as conservatory plants.

Cestrum nocturnum (night jessamine) – a West Indian shrub with lanceolate evergreen leaves, panicles of greenish-white tubular flowers which are powerfully scented at night, and poisonous white berries. John Innes No. 3 with a winter minimum temperature of 45°F (7°C).

Chrysanthemum frutescens see *Argyranthemum frutescens*.

Clethra arborea – an evergreen shrub from Madeira with handsome serrated oblanceolate leaves covered in ginger-brown hairs and 6 in (15 cm) panicles of cup-shaped scented flowers in late summer and early autumn. Lime-free compost with extra peat and sand; winter minimum 45°F (5°C); 10–15 ft (3–4.5 m).

Datura cornigera (angel's trumpet, horn of plenty) – a semi-evergreen Mexican species with downy pointed ovate leaves and heavily scented 6 in (15 cm) pendant trumpets all summer; 'Knightii'* – double white trumpets. Cultivation as above, keeping an eye open for red spider and white fly, feeding regularly when in growth and cutting back hard in early spring; 6–8 ft (1.8–2.4 m).

Datura metel – an annual shrubby species from India with red stems, dark green ovate leaves and 8 in (20 cm) creamy-white flowers in early summer. Needs full sun and John Innes No. 2; 2–4 ft (60 cm–1.2 m).

Datura meteloides – a short-lived shrubby species from Texas with unpleasant-smelling hairy grey foliage and delightfully scented upright 6 in (15 cm) trumpets which are gener-

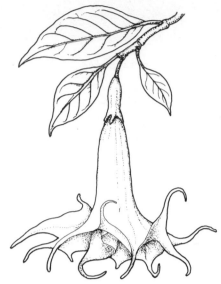

Datura cornigera 'Knightii'

ally white with a tinge of lilac, though on some plants may be more or less bluish-mauve. Cultivation as above, overwinter on the dry side at a minimum 45°F (7°C); 3 ft (90 cm).

*Datura suaveolens** – another Mexican species, bushy and compact with elliptic-oblong leaves with large pendant white trumpets in summer. Cultivation and precautions as above; 10 ft (3 m) or more.

Euphorbia pulcherrima (poinsettia) – a deciduous Mexican shrub with emerald green lobed elliptic leaves and insignificant flowers surrounded by showy 6 in (15 cm) bracts; an excellent winter-flowering pot plant, usually discarded after the first flowering but may be grown on under glass and stood outdoors during summer; 'Alba'* – creamy-white bracts. Peat-based compost or John Innes No 2; enjoys moisture, humidity and feeding when growing; needs bright light all year round, just moist and at a minimum 55°F (13°C) in autumn and winter, increasing to 65°F (18°C) in spring; up to 5 ft (1.5 m).

Fuchsia x *hybrida* – indispensable deciduous or semi-evergreen shrubs for pots, bedding, hanging baskets and for training as standards for tubs; may be treated as annuals or over-wintered at a minimum 40°F (5°C) and grown on to specimen size; easily propagated by cuttings taken from new shoots in spring and inserted in equal parts peat and sand at 60°F (16°C);

'Annabelle' – flouncy double white with pink stamens; 'Constellation' – ivory double flowers; 'F. M. Abbott'* – a dainty single white with a compact habit, excellent for small containers; 'Sleigh Bells' – a large single white with pink veins and stamens; 'This England' – golden foliage and blush-white double flowers; 'Ting-a-ling'* – a free-flowering elegant cultivar with beautifully shaped single white flowers and pink-tipped stamens; 'White Fairy' – double white. Peat-based compost for those treated as annuals, John Innes No. 3 for specimen plants; prone to white fly and red spider mite when grown under glass; height according to habit and age.

Gardenia jasminoides (Cape jasmine)* – an evergreen Chinese species with glossy dark green ovate leaves and heavily scented waxy white flowers, 2–3 in (5–7.5 cm) across; rarely seen in its wild single form; 'Florida' – the summer-flowering double form; 'Veitchiana' – double and winter-flowering. Peaty compost; keep an eagle eye open for red spider mite, aphids, mealy bugs (you name it, gardenias get it); ample warmth, humidity and feeding during the summer; ease up on the water and maintain a winter minimum of 55°F (12°C) for summer-flowerers; ensure at least 60°F (16°C) and continue watering winter-flowerers; 2–4 ft (60 cm–1.2 m). ✕

Helichrysum bracteatum (straw flower) 'Skynet' – a compact long-lived shrubby form of the familiar everlasting flower with papery double cream flowers in summer and autumn. John Innes No. 2 in full sun, frost-free in winter; 2–3 ft (60–90 cm). ✕

Heliotropium x *hybridum* (cherry pie) – evergreen perennial sub-shrubs derived from two summer-flowering Peruvian species, usually grown as half-hardy annuals; wrinkled, slightly hairy oblong-lanceolate leaves and branched corymbs of tiny forget-me-not flowers with a wonderful scent of cooked cherries; very popular in Victorian times and still a great favourite with butterflies; 'White Lady' – a Victorian variety (possibly lost to cultivation with the present vogue for dark purple cultivars) with light green foliage and the palest lilac flowers which fade to white, especially under glass. May be trained as a standard if treated as perennial; John Innes No. 2 in sun or light shade; overwinter at 50°F (10°C) and prune in early spring; 12–18 in (30–45 cm) (more as a standard).

Hibiscus rosa-sinensis – a Chinese species now so widely grown as to be symbolic of exotic places; shiny dark green ovate leaves and a succession of 5 in (12.5 cm) flowers in summer; 'Casablanca' – large single white. Peat-based compost, plenty of light and a winter minimum of 50°F (10°C).

Luculia grandifolia – an evergreen shrub from Bhutan with ovate leaves exceeding 12 in (30 cm) in length and clusters of fragrant pure white jasmine-like flowers, each about 2 in (5 cm) across, in late spring and summer. John Innes No. 3; needs ample moisture, light and ventilation in summer and drier conditions at a minimum 45°F (7°C) in winter; prune after flowering; 10 ft (3 m) or more.

Nerium oleander (oleander) – a Mediterranean shrub with narrow dull green leathery leaves and clusters of showy tubular flowers which open into five flat petal-like lobes; excellent for pots which can be stood outside on a sun-drenched patio in summer; 'Album' – semi-double scented flowers; 'Album Plenum' – large double white rose-like flowers; 'Madonna Grandiflorum' – a very fine large creamy-white double; 'Soeur Agnes' – pure white single. John Innes No. 2; full sun, copious water and ventilation in summer; on the dry side, again in full sun, and at a minimum 45°F (7°C) during winter; needs sun and warmth (55–65°F/13–18°C) in spring for successful flowering; 6 ft (1.8 m) or more.

*Pittosporum tobira** – an Oriental species with shiny obovate leathery leaves and 2–3 in (5–7.5 cm) clusters of cream flowers in spring, whose strong scent resembles orange blossom; 'Variegatum' – creamy-white margins. Sandy peat-based compost, full sun and winter minimum of 40°F (5°C).

Prostanthera cuneata (Australian mint bush) – small, strongly aromatic evergreen leaves and numerous tiny white flowers, minutely purple-spotted in summer. Well-drained peat-based compost with additional sand, on the dry side in winter and at a minimum of 40°F (5°C).

Sparmannia africana (African hemp) – a quick-growing evergreen shrub, easy from seed or cuttings and flowering when young; fresh green toothed heart-shaped leaves covered in short fine hairs and clusters of 1 in (2.5 cm) white flowers in early summer, each with a prominent boss of purple-tipped gol-

Pittosporum tobira

den stamens which open widely when blown or touched; 'Flore Pleno' – double flowers. John Innes No. 3; plenty of sun and water during the summer; minimum 45°F (7°C) in winter; up to 15 ft (4.5 m) but can be cut back drastically in spring.

APPENDIX

White Flowers and their meanings in Victorian florigraphy

alyssum – worth beyond beauty

amaryllis (any colour) – pride, haughtiness

anemone (windflower) – sickness; expectation

angelica – inspiration; magic

asphodel – unending regrets; 'My regrets follow you to the grave'

aster, double (any colour) – 'I share your feelings'

aster, single (any colour) – 'I will think about it'

azalea (any colour) – temperance

basil – hatred

bellflower, white – gratitude

bergamot, wild (*Monarda amplexicaulis*) – 'Your whims are quite unbearable'

bindweed, great – insinuation

bindweed, small – humility

blackthorn – difficulty

bogbean – calm; repose

Brunfelsia latifolia – 'Beware of false friends'

butterfly orchid – gaiety

calla lily – magnificent beauty

camellia, white – perfected loveliness

camomile – courage in adversity

candytuft (any colour) – indifference

catchfly, white – betrayed

cherry, white – good education

chervil – sincerity

chickweed – rendezvous

Christmas rose – 'Relieve my anxiety'

chrysanthemum, white – truth

cineraria (any colour) – always delightful

cistus – security; popularity

clarkia (any colour) – 'The variety of your conversation delights me'

clematis (any colour) – filial love

clematis, evergreen (*C. armandii*) – poverty

clover, white – 'Think of me'; promise

Cobaea (purple or white) – gossip

columbine (any colour) – folly

coriander – hidden worth

crab apple blossom – ill-nature

crowfoot – ingratitude

cyclamen (any colour) – diffidence

dahlia (any colour) – instability; pomp

daisy – innocence

daisy, Michaelmas (any colour) – farewell; cheerfulness; afterthought

Datura (trumpet flower) – fame

Davidia (handkerchief tree) – warning

Diosma ericoides (African steel bush) – 'Your simple elegance charms me'

dittany of Crete, white – passion

dogwood blossom – 'Am I perfectly indifferent to you?'

elder – compassion; zeal

enchanter's nightshade – witchcraft; sorcery

fair maids of France (*Ranunculus aconitifolius*) – lustre

forget-me-not (blue or white) – remember me; forget-me-not; true love

foxglove (purple or white) – insincerity

gardenia – refinement

gillyflower, (pink) (any colour) – bonds of affection

gladiolus (any colour) – ready armed

goat's rue (lilac or white) – reason

guelder rose – winter; age

hawthorn – hope

heather (purple or white) – solitude

hemlock – 'You will be the death of me'

henna (camphire) – fragrance

holly – foresight

hollyhock, white – female ambition

honesty (purple or white) – honesty; fascination

honeysuckle (woodbine) – fraternal love

horse chestnut – luxury

Houstonia caerulea (blue or white) – content

Hoya carnosa – sculpture

Hoya bella – contentment

hyacinth, white – unobtrusive loveliness

hydrangea (any colour) – presumption; a boaster

hyssop (any colour) – cleanliness

iris (any colour) – message

Jacob's ladder (blue or white) – come down

jasmine – amiability

jasmine, Spanish – sensuality

jonquil, white – 'I desire a return of affection'

larkspur (any colour) – lightness; levity

laurel – perfidy

lavender (any colour) – distrust
lemon balm – pleasantry
lemon blossom – fidelity in love
lilac, white – youthful innocence
lily, white – purity; sweetness
lily-of-the-valley – return of happiness
lobelia (any colour) – malevolence
lotus flower – estranged love
love-in-the-mist (blue or white) – perplexity; embarrassment
lupin (any colour) – voraciousness
magnolia – love of nature
Magnolia grandiflora – dignity
meadow saffron (lilac or white) – 'My best days are past'
meadowsweet – uselessness
mock orange – counterfeit
motherwort – secret love
mountain ash (rowan) – prudence; 'With me you are safe'
mouse-ear chickweed – ingenuous simplicity
mullein, white – good nature
myrtle – love
narcissus – egotism
night-blooming cereus – transient beauty
night convolvulus (*Ipomoea bona-nox*) – night
oleander (any colour) – beware
olive – peace
orange blossom – 'Your purity equals your loveliness'; chastity
ox-eye daisy – patience; a token
parsley – festivity
pasque flower (purple or white) – 'You have no claims'
passion flower (upside-down) – superstition; (upward facing) – faith
pea, everlasting (purple or white) – lasting pleasure
pea, sweet (any colour) – departure
pearl everlasting – never-ending remembrance
peony (any colour) – shame; bashfulness
periwinkle, white – tender recollections
phlox (any colour) – unanimity
pink, white – ingeniousness; talent; fair and fascinating
plum, cherry (*Prunus cerasifera*) – privation
poppy, white – sleep; my bane
potato flowers – benevolence
privet – prohibition
rose, white (bud) – a heart ignorant of love; girlhood
rose, white (dried) – death preferable to loss of innocence

rose, white (full-blown) – 'I am worthy of you'
rose, white (withered) – transient impressions
rose, white and red together – unity; reciprocal love
sage (any colour) – esteem; domesticity
saxifrage, mossy – affection (maternal)
scabious (any colour) – unfortunate attachment
Schinus terebinthifolius (Christmas berry tree) – religious fervour
shepherd's purse – 'I offer you my all'
snapdragon (any colour) – presumption; 'No!'
snowball tree – bound
snowdrop – hope
star of Bethlehem (*Ornithogalum umbellatum*) – guidance; purity
stephanotis – 'Will you accompany me to the East?'
stock (any colour) – lasting beauty
stock, ten-week (any colour) – promptness
strawberry flowers – foresight
sweet sultan, white – sweetness
sweet william (any colour) – gallantry
thorn apple (*Datura stramonium*) – deceitful charms
thrift (pink or white) – sympathy
thyme (any colour) – activity; courage
traveller's joy – safety
tuberose – dangerous pleasures; voluptuousness
valerian – an accommodating disposition
verbena, white – 'Pray for me'
violet, white – candour; innocence; modesty
water lily – purity of heart
wisteria (lilac or white) – 'Welcome, fair stranger!'
wood sorrel – joy; maternal tenderness
yarrow – war
Xeranthemum (purple or white) – cheerfulness under adversity
zinnia (any colour) – thoughts of absent friends

Words for white

Botanists have come up with numerous names for white flowers. The scientific names for plants are in Latin and therefore grammatical gender must be observed. For example, the word for 'white' is *alba* when the noun is feminine, *albus* when masculine, and *album* when neuter. In the following list, only the masculine form of the adjective is given.

albicans – being white
albidus – white
albiflorus – white-flowered
albispathus – white-spathed
albulus – whitish
albo-maculatus – white-spotted
albus – white; bright
argyraeus – silvery-white
candicans – being white, bright, radiant, beautiful
candidissimus – whitest; brightest; most radiant; most beautiful
candidus – white; bright; radiant; beautiful
canescens – greyish-white; hoary
chionanthus – white-flowered (from the Greek *chion*, meaning 'snow')
eburneus – ivory-white
Galanthus – snowflower (from the Greek, *gala* – milk; *anthos* – flower)
glacialis – frosty; icy, belonging to snowy regions
lacteus – milk-white
lactiflorus – with milky-white flowers
leucotrichus – white-haired
lucidus – luminous
nitidus – bright; shining
nivalis – snowy
niveus – snow-white
noctiflorus – night-flowering (though not necessarily white)
nocturnus – of the night (comment as above)
pallidus – colourless; pale
tristis – sad; sombre (used of greenish-white)
vestalis – like a virgin priestess
virginalis – virginal

Suppliers with an interesting range of white flowers

MO = mail order
EX = export

AVON BULBS, Upper Westwood, Bradford-on-Avon, Wiltshire BA15 2AT (02216) 3723 (bulbs, especially dwarf and uncommon species) MO/EX

BLOM'S BULBS (Walter Blom & Son Ltd), Coombelands Nurseries, Leavesden, Watford, Hertfordshire WD2 7BH (0923) 672071/673767 (bulbs) MO/EX

BRESSINGHAM GARDENS, Diss, Norfolk IP22 2AB (037988) 464 (shrubs, alpines and herbaceous) MO/EX

BURNHAM'S ORCHIDS, Forches Cross, Newton Abbot, Devon TQ12 6PZ (0626) 52233 (wide range of species and hybrid orchids)

CHESSINGTON NURSERIES LTD, Leatherhead Road, Chessington, Surrey (03727) 25638 (choice house and conservatory plants).

DAVID AUSTIN, Bowling Green Lane, Albrighton, Wolverhampton WV7 3HB (090722) 3931 (roses, irises, peonies, herbaceous plants generally) MO/EX

DE JAGER, Marden, Kent TN12 9BP (0622) 831235/831541) (bulbs, some perennials and aquatics) MO

HILLIER NURSERIES (WINCHESTER) LTD, Ampfield House, Ampfield, Romsey, Hampshire SO51 9PA (0794) 68733 (wide range of trees, shrubs, roses, some hardy perennials) MO

HOLLINGTON NURSERIES, Woolton Hill, Newbury, Berkshire RG15 9XT (0635) 253908 (herbs, conservatory plants) MO

HOLLY GATE CACTUS NURSERY, Billingshurst Road, Ashington, West Sussex RH20 3BA (0903) 892930 (cacti and succulents, including Selenicereus, epiphyllums, Christmas cacti, also geraniums) MO/EX

KELWAYS NURSERIES, Langport, Somerset TA10 9SL (0458) 250521 (bulbs and herbaceous, especially irises, peonies) MO

LANGTHORNES PLANTERY, Little Canfield, nr Dunmow, Essex CM6 1TD (0371) 2611 (choice trees, shrubs and herbaceous)

NEWINGTON NURSERIES, Newington, Oxford OX9 8AW (0865) 891401 (conservatory plants) MO

ORCHIDS BY TERRY ADNAMS, Thatched Lodge, 92 Bury's Bank Road, Crookham Common, Newbury, Berkshire RG15 8DD (0635) 42566 (orchids) MO/EX

OVIATT-HAM. M, Ely House, 15 Green Street, Willingham, Cambridge CB4 5JA (0954) 60481 (clematis, white passion flower and other climbers) MO/EX

RAMPARTS NURSERY, Bakers Lane, Colchester, Essex CO4 5BB (0206) 72050 (herbaceous, especially pinks and grey/silver-leaved plants) MO

STAPLEY WATER GARDENS LTD, 92 London Road, Stapeley, Nantwich, Cheshire CW5 7LH (0270) 623868 (hardy and tropical water lilies, aquatic and marsh plants) MO/EX

THOMPSON AND MORGAN, London Road, Ipswich, Suffolk IP2 0BA (0473) 6888588 (wide range of seeds of white-flowered forms/varieties/species) MO/EX

TOWNSEND, K. J., 17 Valerie Close, St Albans, Hertfordshire AL1 5JD (Achimenes) MO

UNUSUAL PLANTS, Beth Chatto Gardens, Elmstead Market, Colchester, Essex CO7 7BD (020622) 2007 (herbaceous) MO

White gardens to visit in England and Wales

Details of gardens open under the National Gardens Scheme can be obtained from *Gardens of England and Wales Open to the Public*, which is published annually and available from book shops. You are advised to check opening days and times before visiting any garden.

ARTHINGTON MANOR (Mr & Mrs W. Guiness), Arthingworth, nr Market Harborough, Leicestershire (open under the NGS): small white border.

BROOK COTTAGE (Mr & Mrs D. Hodges), Alkerton, nr Banbury, Oxfordshire (open under the NGS and by appointment): white border backed by hedge, fine *Cobaea scandens* 'Alba' on house wall.

CHIVEL FARM (Mr & Mrs. J. D. Sword), Heythrop, nr Chipping Norton, Oxfordshire (open under the NGS): small formal white garden.

COTON LODGE (Mr & Mrs A. de Nobriga), Guilsborough, Northamptonshire (open under the NGS): silver and white garden.

FOLLY FARM (The Hon. Hugh & Mrs Astor), Sulhamstead, Reading, Berkshire (open under the NGS): house by Lutyens and garden by Jekyll; raised white garden.

GREYS COURT (Lady Brunner; The National Trust), Rotherfield Greys, Henley-on-Thames, Oxfordshire (open under the NGS and afternoons Mon–Sat, spring to autumn): walled white garden with central pool, lawn, fine magnolia underplanted with *Allium triquetrum*, *Poncirus* arch, mixed borders.

HAZELBY HOUSE (Mr & Mrs M. J. Lane Fox), North End, Newbury, Berkshire RG15 0AZ (open under the NGS and by written appointment): small well-planted white garden, largely herbaceous with fine *Crambe*, divided into four by paved paths and enclosed by hedge.

HIDCOTE MANOR GARDEN (The National Trust), Mickleton, nr Chipping Campden, Gloucestershire (open daily except Tues and Fri): formal white garden enclosed by hedges and divided into four by paved paths; superb *Lilium martagon* 'Album'.

MANOR HOUSE FARM (Mr & Mrs J. M. Clissold), Nortoft, Guilsborough, Northamptonshire (open under the NGS): silver and white garden.

NEWBY HALL (R. E. J. Compton Esq), Ripon, Yorkshire (open Tues–Sun, spring to autumn): white border backed by hedge.

NINGWOOD MANOR (Lt-Col & Mrs K. J. Shapland), Shalfleet, nr Yarmouth, Isle of Wight (open under the NGS): formal white garden.

PARC GWYNNE (Mr & Mrs W. Windham), Boughrood, nr Brecon, Powys (open under NGS): riverside garden with white and apricot borders.

PEOVER HALL (Mr & Mrs Randle Brooks), Over Peover, Knutsford, Cheshire (open under the NGS; Mons excluding bank holidays and Thurs afternoons, and by appointment, late spring to autumn): walled white and pink gardens, white and blue borders.

13 SELWOOD PLACE (Mrs Anthony Crossley), South Kensington, London SW7 (open under the NGS): green and white border.

SISSINGHURST CASTLE GARDEN Cranbrook, Kent (Nigel Nicolson Esq; The National Trust) (open afternoons Tues–Fri, all day Sat/Sun, spring to autumn): large enclosed white garden with central rose arbour, superbly planted and divided into four by paved paths.

STONEWALLS (Mr & Mrs M. Adams), nr Deddington, Oxfordshire (open under the NGS): white border and excellent clematis collection.

SUTTON BONINGTON HALL (Lady Elton), Sutton Bonington, nr Loughborough, Leicestershire (open under the NGS): formal white garden.

WATERCROFT (Sir Barrie & Lady Heath), Penn, nr Beaconsfield, Buckinghamshire (open under the NGS and by appointment): organic garden on clay with special interest in white flowers.

WOODPECKERS (Dr & Mrs A. J. Cox), Marlcliff, nr Bidford-on-Avon, Warwickshire (open under the NGS): white garden within 2½-acre informal garden.

YORK GATE (Mrs Sybil B. Spencer), Back Church Lane, Adel, Leeds (open under the NGS): small white and silver garden within 1-acre garden.

Index

Note: Bold figures denote photograph or illustration.

Acknowledgements

I have been writing this book for only a year, but in some ways it began years ago. The love of plants and gardening began in early childhood with my parents' and grandparents' gardens, and especially with my grandfather who showed me tigridias and zinnias, lilies and parrot tulips, and left me wide-eyed at their beauty. Next, the works of Dante, William Wordsworth and T. S. Eliot, and the ideas of John Ruskin left a lasting impression during my student days at Manchester University, during which time I also owed a great deal to my tutor, Dr Arnold Goldman, who inspired me to explore images. In more recent years, as my interest grew in natural history generally and nature photography in particular, I came under the spell of naturalists such as Henry Thoreau, Reginald Farrer, Peter Matthiessen and Eliot Porter. All these and others too numerous to mention have instilled in me, drop by drop over the years, a curiosity about the plant world which now spans botany, ecology, horticulture, medicine, food, art, history and religion. Quite where the fascination with white flowers came from, I don't know. Perhaps they are just my own particular lunacy, though an astrologer tells me that an interest in white flowers, and especially in night-flowering species may be accounted for by the fact that the moon looms large in my horoscope!

But I must stop short of acknowledging the pervasive influence of the moon when there are so many more immediate debts of gratitude! Although inspiration and influences are fundamental to a book, it is the nuts and bolts of research, illustration, editing, design and production which ensure that it emerges, fully formed, from the dark – if moonlit – recesses of the imagination. This process has been helped by many, most of all by the enthusiasm and skills of my editors Connie Austen-Smith, Clare Ford and Lesley Gowers, illustrator John Wilkinson and designer Julian Holland. I must also thank the Director and librarians of the Royal Botanic Gardens, Kew, and the librarians of the Lindley Library for their unfailing helpfulness, and all those whose plants and gardens I photographed – especially Mr and Mrs M. Lane Fox whose garden is such a delight. Lastly, heartfelt thanks to family and friends for their generous support which keeps the cogs turning. Above all, love to my children Anna, William and Robin, who share my life and hopes as no one else can.